Living the Dream

Cognitive Dream Interface Theory (CDIT), a Philosophical and Scientific Inquiry into Consciousness and Reality

Living the Dream

Cognitive Dream Interface Theory (CDIT), a Philosophical and Scientific Inquiry into Consciousness and Reality

by Andrew T. Jaffe, M.D.

Essentia

London, UK
Washington, DC, USA

CollectiveInk

First published by Essentia Books, 2026
Essentia Books is an imprint of Collective Ink Ltd.,
Unit 11, Shepperton House, 89 Shepperton Road, London, N1 3DF
office@collectiveinkbooks.com
www.collectiveinkbooks.com
www.essentia-books.com

For distributor details and how to order, please visit the 'Ordering' section on our website.

ISBN: 978 1 917991 62 9
978 1 917991 72 8 (ebook)
Library of Congress Control Number: 2025945465

www.cdit-livingthedream.com
A CIP catalogue record for this book is available from the British Library.

Design: Lapiz Digital Services

UK: Printed and bound by CPI Group (UK) Ltd, Croydon, CR0 4YY
Printed in North America by CPI GPS partners

The manufacturer's authorised representative in the EU for product safety is:
eucomply OÜ - Pärnu mnt 139b-14, 11317 Tallinn, Estonia, hello@ eucompliancepartner.com,
www.eucompliancepartner.com

Table of Contents

Preface

Studying biology in college and then medicine, eventually becoming a physician, you could say I've been a card-carrying materialist for most of my life. The scientific method was my compass, and I saw the world through the lens of cause and effect, chemistry and cells. But over the last decade, through exploration of my own mind via meditation and a growing fascination with the enigma of consciousness, cracks began to form in my materialist wall.

The Dreamer, the Dream, and the Dream Character

This book began as a collection of thoughts, questions, and observations I couldn't shake; late-night realizations, quiet reflections, and the sense that something about our experience of being alive is far more mysterious than we're often led to believe. Like many, I've felt the strangeness of existence and wondered: Why am I aware at all? Where did this come from? And what is this world I find myself in?

In trying to answer those questions, I found guidance and inspiration from several modern thinkers whose work challenged me to look deeper, not just outward into the universe, but inward into the nature of consciousness itself.

Sam Harris, through his book *Waking Up* and his meditation guidance, helped shine a light on the inner landscape of consciousness. His repeated invitation to "look for the one who is looking" became a pivotal pointer in my own journey, a simple, powerful method of stripping away illusion to touch the essence of awareness itself.

Rupert Spira, with his lucid and poetic teachings, offered the language and structure to recognize awareness as the ever-present background of experience. His use of the dream metaphor, not as mere imagery, but as a philosophical framework deeply resonated with me and inspired me to follow that analogy further than tradition often dares.

Bernardo Kastrup, with his scientific rigor and contemporary defense of idealism, provided a bridge between philosophy, science, and consciousness. His willingness to challenge materialist assumptions and offer logically sound alternatives gave me the confidence to formulate and develop the model presented here.

I'm also deeply grateful for the ongoing public discourse that helped shape my thinking. Several podcast platforms have served as modern-day salons for intellectual exploration, exposing me to a range of perspectives that challenged and sharpened my views. In particular, *Theories of Everything* with Curt Jaimungal, *The Michael Shermer Show*, Brian Keating's *Into the Impossible*, and Sean Carroll's *Mindscape* offered invaluable conversations that illuminated both the frontiers of physics and the enduring questions of consciousness. These dialogues didn't dictate the direction of this book, but they helped cultivate the landscape from which my own theory would emerge.

While the ideas in this book are my own interpretation, they are deeply indebted to these thinkers and many others. Each, in their own way, pointed me to the same ineffable conclusion: that awareness is not a byproduct of the world, but the ground from which the world arises.

This work expands on that insight by offering a coherent framework that blends idealism, science, and the dream analogy into a single thesis: that we are not simply in the universe, we are the universe dreaming itself, and awareness is the dreamer. The world is its dream.

I call this framework **Cognitive Dream Interface Theory**, or **CDIT**. CDIT proposes that what we take to be physical reality is a symbolic interface rendered within awareness, not by a brain, not by a god, and not by a machine, but by awareness itself as a spontaneous, natural phenomenon. Just as dreams at night can generate immersive realities with no external input, so too might waking life be the projection of a deeper cognitive field we mistake for the world.

While this book is fundamentally philosophical and experiential in its foundation, I also engage with modern science, especially quantum physics and biology, as circumstantial evidence in support of the theory. I make a deliberate effort not to overstate speculative interpretations or bend ambiguous data to fit my model. Rather, I point to places where mainstream science, already struggling with its own paradoxes, begins to align with the implications of CDIT. These intersections do not serve as proof, but as signposts: indicators that an idealist view of reality may be not only coherent, but more elegant and explanatory than its materialist counterparts.

Chapter 1

The Mystery of Being

You find yourself here. You're not entirely sure how, or why, but here you are.

The Great Reversal

From a materialist perspective, this is a bewildering proposition. Out of oblivion, of infinite nothingness, you somehow emerge as a sentient being for a mere 85 years if you're lucky, only to vanish again into the same silent void. The odds of existing now, at this exact time and place in the cosmos are astronomically low, virtually zero, if existence is just a random arrangement of particles in an indifferent universe. And yet, here you are.

This, in itself, should cast doubt on whether the materialist framework is the right starting point. It may explain mechanics, but it struggles to explain the miracle of being. The raw fact of awareness seems far more certain than any assumption about particles, space, or time.

The Illusion of a World Outside

There's a world, a body, a past, at least, that's the story you've inherited. But if you pause for just a moment and peel away everything you've been told, everything you assume, everything you think you know, what remains? What can you be absolutely certain of?

Only this: you are aware.

This may seem like a small or obvious point, but it's neither. It's the foundation of everything. You are aware, not necessarily aware of who you are, where you are, or even what you are, but aware nonetheless. That awareness is not something you do; it is something you are. It precedes your thoughts, outlasts your emotions, and quietly witnesses the parade of sensations that make up what you call your life.

Everything else, your memories, your identity, the so-called outside world is part of the content of awareness. But awareness itself? It never appears as content. You cannot find it the way you find objects in space. You cannot see it like you see a tree or measure it like a heartbeat. It's not an object within your field of experience. It is the **presence beneath the field**, the silent backdrop in which all appearances unfold.

And what's most striking is that this awareness doesn't age. It doesn't drift. It doesn't evolve. The body does, the personality does, even the brain changes over time, but awareness remains pristine. The same witnessing presence that looked out from behind your eyes at five years old looks out still. It is unmodified by the passing of time. It does not remember, though memory appears within it. It does not anticipate, though anticipation arises inside it. It simply is.

This has radical implications for what we consider real. If awareness is the only undeniable fact, the only thing that cannot be doubted, then perhaps it is not simply one thing among many. Perhaps it is the foundation. Perhaps everything else; time, space, matter, causality emerges not as fundamental features of the universe, but as appearances within awareness itself.

This flips our traditional worldview. In the standard materialist view, consciousness is a latecomer in a vast and ancient cosmos, a byproduct of brains, which are themselves products of matter, born from stars, born from a Big Bang. But in the view we're exploring here, a view rooted in idealism, awareness is not the product of the universe. **The universe is the product of awareness.**

This isn't as outlandish as it sounds. Even mainstream physics admits that our access to the so-called external world is entirely mediated by perception. You never directly experience a photon or an atom or a quark. You experience colors, shapes, sounds;

all interpreted and assembled in the brain and presented to consciousness (Hoffman, 2019). But what if consciousness isn't just the passive recipient of this information? What if it's the stage on which all of it appears?

The implications are vast.

In dreams, a complete world can arise; people, places, sensations, even a sense of time and identity, and yet it all emerges from the mind of the dreamer. Nothing in the dream exists independently of that mind.

The dream feels real while it's happening, and only upon waking do we say, "It was just a dream."

But what if waking life is also a dream, a consistent, structured, communal dream taking place within a larger mind? What if that larger mind is the same awareness you call "I"?

That would mean the world doesn't exist apart from awareness. It exists within it. It would mean you're not a creature in the universe, **you are the universe, dreaming itself into form, forgetting its true nature in order to remember it anew**. Your existence, then, was not a statistical fluke, but inevitable, as certain as you appearing in your own dreams at night. Materialism offers no reason why you, this exact viewpoint, should exist rather than any other. Under CDIT, it's not a miracle or a fluke. It's the natural coherence of awareness dreaming itself into form.

This isn't spiritual metaphor. This is philosophical reasoning, and, increasingly, scientific reasoning. Quantum physics has already shown that measurement and/or observation may play a fundamental role in the behavior of matter (Bell, 1964). Time and space themselves may not be fundamental (Barbour, 1999; Rovelli, 2022). The Big Bang may not be the beginning. Information may be more basic than matter (Wheeler, 1990). And all of it seems to hinge on the role of awareness.

You didn't choose to be here, in this moment, reading these words. You simply find yourself present, an observer within an unfolding experience. This experience may appear vividly real, but there is no scientific proof that any interaction is objectively occurring between a separate subject and object in physical space. The only certainty is that something is happening, and that you are aware of it. That awareness remains unchanged across waking life, dreams, and altered states; **it is the constant witness.**

Enter CDIT

In the chapters ahead, we'll explore a framework called Cognitive Dream Interface Theory (CDIT), which builds on this idea by proposing that our experience of reality is a dream-like interface rendered by awareness. In CDIT, brains, bodies, and environments are not literal objects existing independently, but symbolic renderings, structured appearances within the dream designed to simulate embodiment, continuity, and separation.

These symbols don't mirror an objective world; they create a coherent one. Just as in a dream, where the scenery is immersive but generated from within, CDIT suggests that what we call "reality" is a deeply consistent, meaningful projection within awareness.

And while the contents of the dream may be symbolic, the awareness of the experience, whether in waking life or dreaming, is the same: unchanged, continuous, undeniably real.

So, this is where we begin. Not with answers, but with the only fact that cannot be denied: **You are aware.**

Everything else is the dream.

Chapter 2

Awareness Comes First

The Central Mystery of Mind

Despite centuries of scientific progress, the question of consciousness remains one of the most puzzling enigmas in both philosophy and neuroscience. What is this mysterious subjective presence that accompanies our experience? How can something so central to our lives, the sense of being aware, still elude objective explanation?

If we begin from the premise that awareness is primary, we find ourselves re-evaluating the very framework through which we interpret reality. The standard model suggests that consciousness arises from physical processes; that neurons fire, signals transmit, and somehow, somewhere along the way, an inner world of experience lights up. But no matter how detailed our maps of the brain become, the leap from matter to experience remains unbridged.

This is what philosopher David Chalmers famously termed **the "hard problem" of consciousness** (Chalmers, 1996). Unlike the so-called "easy problems" such as explaining how the brain integrates information, discriminates stimuli, or produces behavior, the hard problem asks: why should any of that produce experience at all? Why should brain activity feel like something from the inside? Why does seeing red or feeling pain come with a qualitative, first-person dimension?

No purely mechanistic account has yet explained why certain physical states should give rise to subjective awareness, rather than just function like input-output machines. It's not a question of complexity; it's a question of ontology. How do we account for the fact that experience exists at all?

Idealism and CDIT: A Unified Interface Model

Idealism offers a radically different perspective: that consciousness is not a latecomer or emergent property, but the **foundational substance of reality**. In this view, the universe does not generate awareness, rather, awareness gives rise to the universe. What we call "matter" may simply be how awareness presents itself within a dream-like projection.

Cognitive Dream Interface Theory (CDIT) proposes that our experience of the world is not of a physical reality "out there," but of a **symbolic dream** projected by awareness itself. In this model, what we call a "brain" is not the source of consciousness, but a symbolic filter constructed within the dream to simulate localization and embodiment. CDIT integrates idealism with the dream analogy to offer a unified explanation of perception, identity, and experience.

This flips the materialist view on its head. The cosmos, under this lens, is not a vast, indifferent expanse of matter evolving blindly through time. It is, at root, **a field of awareness expressing itself through evolving forms.**

Hoffman's User Interface Theory

This is not just metaphysical speculation. Some of the most provocative modern scientific theories echo this inversion. Donald Hoffman, a cognitive scientist, argues in his **User Interface Theory** that our perception is not a transparent window to reality but a symbolic interface, much like icons on a desktop (Hoffman, 2019). We do not see the world as it is, but as it is useful to us. Evolution, Hoffman argues, favored fitness over truth. Our senses evolved not to reveal objective reality, but to help us survive, which means we may be interacting with something profoundly real, but perceiving it through an illusion optimized for functionality.

From this angle, consciousness is not something created by the brain, but something the brain accesses or filters. The brain may be more like a transceiver, a tuning device for awareness, than a generator of it. Another way to think about it is like a television set: the TV produces images and sounds, but the content is not inside the machine. It is broadcast from elsewhere. Nor is the television itself conscious of the show. The awareness, the observer, sits apart, watching the transmission from a distance. Similarly, in CDIT, the brain creates the symbolic structure of perception, but the knowing of experience, the awareness itself, is not contained within the apparatus. It witnesses the show from beyond.

Quantum Physics and the Observer

This reframe gains further traction in quantum physics. The **observer effect** posits that the mere act of observation changes the outcome of quantum systems (Wheeler, 1990). Particles do not appear to settle into definite states until measured. Whether this collapse is due to the measuring apparatus or the presence of an observer, the deeper point is that quantum events are not fully determinate without an observation-like interaction. From a CDIT perspective, this suggests that the world is not a pre-existing, fixed structure "out there," but a rendering that becomes coherent only when required for experience.

Bell's Theorem, introduced by physicist John Bell in 1964, posed a direct challenge to classical assumptions about the universe (Bell, 1964). By demonstrating that no theory based on local hidden variables could replicate the predictions of quantum mechanics, Bell showed that entangled particles defy explanation through conventional, cause-and-effect reasoning. Subsequent experiments have upheld his inequality violations, confirming that the behavior of one particle can be instantly correlated with another, no matter the distance.

This leaves us with two striking possibilities: either reality is fundamentally non-local, defying classical cause-and-effect, or measurement itself is integral to how outcomes are rendered.

In both cases, the notion of a mind-independent, objective reality begins to dissolve.

From an idealist view, this is not a bug in the quantum system but a clue to its nature. If awareness is primary, then the fact that physical events require observation-like conditions to take definite form makes sense. There is no universe "out there" fully independent of awareness; what we encounter is always already rendered through the field of consciousness.

Adlam and the Collapse of Linear Time

The implications become even more profound when viewed alongside the work of physicist **Emily Adlam**. In her "all-at-once" interpretation of quantum mechanics, Adlam proposes that the universe is not evolving forward through time but is instead governed by a **final boundary condition** (Adlam, 2014). In other words, the whole universe, past, present, and future exists as a complete configuration, with events co-determined by both initial and final states. This model supports a form of **temporal non-locality**, where future outcomes can influence the past.

Such a structure is difficult to reconcile with materialist intuitions, but it aligns naturally with the idea that time is not fundamental. If time is a dreamlike projection, a mode of conscious experience, then Adlam's framework makes perfect sense. Consciousness is not riding a timeline. It is the space in which the illusion of time unfolds.

Consciousness as the Unchanging Ground

In every state, waking, dreaming, meditative, or anesthetized, awareness remains unaffected. The contents of consciousness change. Stories shift. Sensations rise and fall. But the **field in which they unfold stays the same**. Awareness is not subject to time. It does not age. It does not begin or end. It is always now.

This view reframes identity, too. Even the sense of self, the coherent "I" who supposedly navigates experience, is a mental construct. Neuroscience shows that the brain stitches this "self" together from various neural subsystems. It can fracture or dissolve entirely. But who witnesses that dissolution? Who sees the self arise and pass away?

The witness, the awareness, never disappears.

Beyond Solipsism

This is not a claim that you, as the individual ego, are creating the world. It is not solipsism. Instead, it proposes that the awareness in which your ego arises, the very awareness reading these words, is the same awareness behind every face, every mind, every life.

The dreamer, or dreaming force behind the dream is singular, but it expresses itself through a diversity of masks.

Conclusion: The Real Inversion

The real mystery is not how matter gave rise to mind. It's how we ever came to believe it did. Perhaps we believed the dream too well. But with awareness as the starting point, the puzzle pieces begin to reassemble.

And what they reveal is not a universe of things, but a **universe of experience** held in the open, timeless, and inseparable field of consciousness.

Chapter 3

Without Form

The Search for the Self

In every human life, there comes a moment, sometimes in stillness, sometimes in crisis, when we realize that the story of who we are doesn't quite explain what we are. We can describe our roles, our preferences, even our memories, but we cannot pin down the one who is aware of them. When we ask, *Who is aware?* the answer is never an object. It's not the body. It's not the thoughts. It's not even the feeling of self.

What we find, instead, is a kind of formless awareness, something that can't be seen but sees everything; something that doesn't move but in which everything moves. This awareness cannot be reduced to form because it has no structure, no boundaries, no weight or dimension. It is not a thing in the world. It is that within which the world appears.

You don't see awareness with your eyes; you see *with* awareness. You don't feel it with your skin; you feel *through* it. It is more intimate than your heartbeat and more constant than your personality. You can live an entire life without noticing it, and yet it is what makes life livable.

Naming What Cannot Be Named

Most traditions have names for this formless aspect of experience; the ground of being, the witness, the selfless self. But naming it does not capture it. In fact, the very act of naming introduces the illusion of separation. What you are is not separate from awareness. *You are awareness,* temporarily identified with a particular body and mind.

When all mental activity quiets, in deep meditation, sensory deprivation, or altered states, what remains is not nothingness, but pure presence. Awareness without content. A clarity that is not made of thought, but that allows thought to appear. This has been consistently observed across contemplative traditions, and increasingly explored in cognitive science through studies of non-dual states of consciousness (Josipovic, 2014).

Consider the experience of being in a sensory deprivation tank. No light, no sound, no motion. The body disappears. Time fades. And yet, something remains, the quiet hum of being. Awareness doesn't vanish; it becomes more obvious. Without the usual distractions, it reveals its true nature: open, boundless, ever-ready to perceive.

This points to a critical distinction: awareness is not what is perceived, but the capacity for perception itself. And it never takes form. Whatever form arises, a thought, a sensation, an object, appears and disappears. But awareness remains unchanged.

Even when our sense of self dissolves, as it can in deep flow states or certain neuropsychological conditions, awareness remains. It's not the self that knows. It's not the ego that perceives. The ego is just another construct, a useful fiction, perhaps, but a fiction nonetheless. The real "knower" is never found among the contents of consciousness. It is the container, the background.

The Neuroscience of the Formless Witness

This insight is echoed, though often unintentionally, in modern neuroscience. Research shows that what we call the "self" is distributed across various brain regions, with no single command center (Northoff & Panksepp, 2008). The sense of "I" is stitched together moment by moment, like a narrative being continuously updated. But the one who *witnesses* that narrative? That remains mysteriously beyond reach.

CDIT interprets this neuroscientific ambiguity through a symbolic lens: the "self" is a dynamically rendered avatar within the dream, constructed by the brain-interface to simulate a localized experiencer. The witness, pure awareness, isn't inside the body. The body is inside the awareness.

If awareness were emergent from brain activity, it should fluctuate fundamentally when the brain changes. But it doesn't. The *content* of consciousness, clarity, mood, sense of self, may change. But awareness itself, the basic, formless presence, remains constant. It is present through dreams, bliss and confusion. If its narrative produces periods of blackout or general anesthesia, those are only symbolic representations within the dream it creates, one's first-person experience is still continuous, without gaps.

Awareness is the only aspect of your experience that is *not* an experience. It does not move with thought. It does not vanish with memory. It is always now.

We are accustomed to identifying with form: bodies, emotions, stories. But form is transient. It emerges, shifts, and dissolves. Awareness doesn't. It is the invisible stage on which the entire play of life unfolds.

To recognize this is not to deny the beauty of form, but to put it in context. Just as the sky does not cling to the clouds that pass through it, awareness does not cling to the thoughts, feelings, or sensations that arise within it.

Awareness allows everything, resists nothing, and quietly abides, untouched by the dramas it contains.

This is not a metaphysical claim but a phenomenological one. You can verify it in your own experience. Before the next thought arises, before the next interpretation, there is already awareness. It is not a belief. It is the fact of being.

And that awareness is not a thing. It has no color, shape, size, or boundary. It cannot be seen, but it is what allows seeing to occur.

You Are Awareness Without Form

In this way, you are not a thing in the world. *You are the world appearing to itself through a temporary lens.* You are not the contents of awareness; you are awareness itself; timeless, centerless, and whole.

Chapter 4

The Brain as Interface

The Hard Problem and a New Perspective

The relationship between the brain and consciousness has long been a central puzzle of neuroscience and philosophy. The dominant narrative suggests that the brain produces consciousness; that somewhere amid the electrical signals and synaptic exchanges, experience arises.

But despite the sophistication of modern brain imaging and computational models, the leap from objective neural activity to subjective awareness remains unexplained. As Chalmers's "hard problem" contends, none of that tells us why red looks red, why pain hurts, or why music moves us.

From the idealist perspective, the brain is not the generator of consciousness but its interface; a receiver, filter, or transducer that allows consciousness to engage with a particular kind of reality. Just as a radio does not create music but tunes into signals already in the air, the brain may shape, localize, and constrain awareness without being its origin.

Consciousness Without Continuity

This framing radically shifts the implications of neuroscience. It suggests that changes in brain function do not produce changes in consciousness per se, but in how consciousness experiences.

Damage to the visual cortex doesn't eliminate consciousness; it changes what consciousness can see.

This sheds light on why unconscious states like deep sleep, general anesthesia, or periods under sedation, do not feel like gaps in awareness from the inside.

While an outside observer may describe these states as blank intervals, the one undergoing them experiences no interruption.

From the perspective of awareness itself, nothing has been lost.

The Filtering Brain and Expanded Awareness

The same reasoning can be extended to altered states. Psychedelics, for example, often reduce activity in the brain's default mode network (DMN), the area associated with the sense of self.

This reduction correlates with experiences of ego dissolution and expanded perception.

Rather than "enhancing" brain activity, these substances often suppress habitual patterns, allowing new modes of awareness to emerge.

Michael Levin's research into bioelectric signaling adds further intrigue. His work shows that even simple organisms, ones without complex brains, exhibit goal-directed, adaptive behaviors, coordinated through electrical fields.

Life's intelligence, it seems, is not purely neuronal but rooted in a deeper morphogenetic order.

The Binding Problem: Fragmented Signals, Unified Experience

One of neuroscience's most enduring puzzles is the *binding problem,* how discrete elements of perception, processed in different parts of the brain, are unified into a coherent, singular experience. Color, shape, motion, and sound are handled by distinct neural circuits, often in separate cortical regions and at varying processing speeds. Yet when we see a red bird in flight, we don't perceive red, motion, and wings as independent data streams, we see a single, meaningful entity: the bird.

This unified field of perception is taken for granted, but it has no fully accepted neurological explanation. Some theories propose synchrony of neural firing, such as gamma wave oscillations, as a solution (Engel & Singer, 2001), while others point to attentional focus as the mechanism (Treisman, 1996). Yet these accounts remain speculative. They describe correlations, not causes. They do not explain how disparate sensory inputs are experienced by a single subject or why there is *any* subjective experience at all.

CDIT reframes this entirely. There is no need to bind separate signals because nothing is truly separate. The dream character appears localized, but the experiencer is awareness itself, already unified, already whole. Perceptual coherence is not something constructed by neurons; it is rendered by awareness through symbolic interface. The "red bird" is not stitched together by brain activity; it is dreamed whole. Neural signatures may accompany the appearance but they are secondary to the dream-like rendering.

This interpretation is consistent with findings in split-brain research. In individuals whose corpus callosum has been severed, their hemispheres function semi-independently yet subjects still report a unified field of awareness (Gazzaniga, 2000). Despite physical division, experience remains coherent. The binding, then, cannot be purely neural, it is a property of awareness.

CDIT proposes that all perceived unity, between senses, between moments, between self and world, emerges not from biological convergence but from the inherently cohesive nature of dreaming awareness. In the same way a sleeping mind effortlessly creates integrated dream scenes from scattered memory and fantasy, waking life is rendered as a singular experiential stream. The brain's role is symbolic, not causal. It marks where attention localizes, but not how unity is achieved.

The mystery of binding fades when we realize the question is not how separate parts come together, but how separateness itself is simulated so convincingly.

Dreaming Brains and Symbolic Filters

Some may ask: if this is a dream, and all dream characters are creations of the dreaming awareness, do those characters actually use their brains as filters?

After all, in dreams, the dreamer's mind generates everything; people, thoughts, environments, without needing internal processing units inside each character.

So how can the brain-as-interface model still apply?

The answer lies in symbolic coherence.

Just as dreams construct backstories and logical continuity on the fly, they also render internal models of functioning bodies and brains.

These symbolic brains aren't necessary for dreaming to occur but simulate filtering to maintain the illusion of individuality.

Thus, in CDIT, brains are not literal filters but symbolic constructs, narrative devices that enable awareness to experience localized, coherent perspectives within the broader dream.

The brain, in this model, is the rendered result of a larger dream-narrative, one in which evolutionary biology provides the internal logic by which form, memory, and behavior appear consistent.

Awareness, seeking immersive variety, dreams not just characters but plausible vehicles for experience, embedding the appearance of brains to simulate individuality.

Evolution is the backfilled structure that gives the dream weight, coherence, and the illusion of a past, not because awareness needs biology, but because the dream requires stability.

Without such coherence, the dream collapses into fractured hallucination.

CDIT presents awareness not as a divine being or metaphysical overseer, but as the natural, foundational condition of existence, the base layer from which all phenomena arise.

Just as gravity doesn't "want" to pull objects together but does so as part of its intrinsic nature, awareness doesn't "desire" to create, rather, its nature is to express itself through the spontaneous unfolding of experiential forms.

We know this because we ourselves are the product of that unfolding; awareness dreaming experience into being.

We see this mirrored every night: our minds, without instruction or design, generate rich, symbolic dreamworlds complete with internal logic and narrative backfill.

CDIT proposes that reality itself functions the same way, not constructed from the outside in, but dreamt from the inside out.

The Observer Beyond the Interface

This perspective also helps explain how one can maintain a persistent sense of self through dramatic physical and psychological changes.

The contents of mind, memories, beliefs, sensations all evolve over time, but the witnessing presence behind those contents remains constant.

We do not feel as though we are a different awareness at fifty than we were at five.

The interface has changed. The perspective has matured.

But the fundamental "I" who knows experience remains unchanged.

In dreams, this becomes even more apparent.

A full world emerges, complete with a dream-body, dream-logic, and dream-memories, all convincingly simulated.

And yet, all of it arises within a deeper awareness that remains constant, even when the dream-self is confused, frightened, or lost.

In the waking state, we believe the brain is "in here" and the world is "out there."

But from the standpoint of awareness, both brain and world are appearances within it.

If we want to understand consciousness, we must not only study the brain.

We must also examine the structure of experience itself; turning the lens inward as lucid dreamers within the dream of waking life.

To do so is to finally consider the possibility that the observer, the one reading these words, is not a creation of the brain, but the very presence in which the brain itself arises.

CDIT frames matter as a stable visual metaphor, the dream's structured texture.

Just like recurring scenery in a dream, matter appears continuous and solid, not because it exists independently, but because awareness renders it consistently for narrative immersion.

Chapter 5

The Illusion of Matter

Matter as a Stable Illusion

When we think of reality, we think of things such as tables and trees, buildings and planets, atoms and quarks. Matter feels like the bedrock of existence: solid, measurable, objective. Consciousness, by contrast, is seen as fragile and fleeting, a thin film atop this dense and enduring foundation.

But what if this intuition is backwards?

Physics has long been revealing that matter is not as solid or fundamental as it seems. At the quantum level, particles do not exist as static things, but as probabilities with tendencies to manifest in particular ways under specific conditions. The famous double-slit experiment demonstrates that particles such as electrons behave like waves when unmeasured and like particles when measured or observed. Some interpretations suggest that observation, though still debated in terms of mechanism, plays a role in how reality collapses into form (Wigner, 1967; Bell, 1964). This raises a profound possibility: that consciousness is not a passive observer, but a participant in the unfolding of physical reality.

Materialist theories have tried to preserve objectivity by proposing mechanisms like the Many Worlds hypothesis, which multiplies entire universes to account for quantum probabilities (Carroll, 2019). But **CDIT** (Cognitive Dream Interface Theory) offers a simpler view: there is one field of awareness, and within it, the apparent solidity of matter is rendered symbolically, much like coherent images in a dream.

Dreamlike Coherence

In dreams, objects feel real. We walk through streets, touch surfaces, taste food. The dream follows rules: gravity, light, and time, until we wake and realize it was all generated within a mind. Matter in the dream was never "made of stuff." It was structure without substance, form without external source.

CDIT proposes that waking life is no different in kind, only in consistency. Matter is rendered as part of a structured symbolic interface. It feels real and obeys laws because the dream is the simulation of embodied experience and demands coherence. The brain, like the hand or the city, is not real in itself; it is a coherent symbol in a shared, immersive dream designed for local experience.

The Interface, Not the Substance

We never directly encounter the "thing-in-itself." What we perceive are rendered interpretations: color, texture, mass, shaped by the cognitive interface of our nervous system. Cognitive scientist Donald Hoffman argues that evolution did not shape us to see objective truth, but to perceive a symbolic interface that favors survival over accuracy (Hoffman, 2019).

CDIT expands this principle: the interface we navigate is not only symbolic, but dream-rendered. Reality is not a neutral structure perceived inaccurately, but a dynamic narrative generated by awareness itself. This awareness is not supernatural, nor intentional. It functions like gravity or magnetism: a natural entity whose very nature is to generate form. Experience unfolds not because it is willed, but because awareness dreams spontaneously, immersively, and with coherent depth.

Matter as Metaphor

From this perspective, matter is a metaphor, a consistent, law-bound expression designed to ground the dream. It allows for learning, effort, growth, physics, pain, joy. It gives the dream gravity and depth. But it is not fundamental. Like a city that appears mid-dream with believable backstory, the universe includes backfilled narratives: the Big Bang, evolution, stars, atoms.

These are not false, rather, symbolic truths. The universe had to appear as if it had always been there. The dream needed a past to make the present coherent. Evolution, chemistry, and physics are the internal logic that gives the dream structure. But the container that dreams them, awareness, is outside time, always now.

The Substance of Awareness and the Dream of Energy

In **CDIT**, awareness is not composed of matter or energy as we traditionally understand them. It is not a thing among things, but the foundational presence from which all things appear. Just as dreams at night arise vividly and spontaneously without any material substrate, the waking world is rendered within awareness as a symbolic simulation.

Energy, light, and matter are dream-textures and coherent appearances shaped by the intrinsic nature of awareness to generate experience. This perspective finds some resonance with the holographic principle in physics, suggesting that our reality may be more informational than material (Susskind, 1995; Bousso, 2002). From the **CDIT** standpoint, light is not the base of existence, rather, it is a medium of symbolic exchange within the dream interface.

Thus, waking life, like the dreams of sleep, is a projection of symbolic experience; rich, lawful, and immersive, but ultimately arising from the timeless, formless presence of awareness itself.

Physical Laws as Coherence Rules

Modern physics increasingly describes reality not as a collection of self-existing substances, but as a system governed by limits, constraints, and relational structures. What once appeared to be fundamental forces or hidden ingredients now appear, upon closer inspection, as rule-like conditions required for the stability of a shared world. Gravity is no longer treated as a force acting between objects but as a geometric feature of spacetime itself. Dark matter is not directly observed as substance but inferred because, without it, large-scale structures would fail to remain coherent. Dark energy is not detected as an entity but named to describe a global tendency of space to expand. Even the speed of light functions not as a property of light alone, but as a universal limit on information and causation. The second law of thermodynamics similarly governs not what exists, but how states unfold over time.

From a CDIT perspective, this convergence is revealing. These phenomena behave less like physical ingredients and more like rules required to sustain a coherent experiential world.

In dreams, awareness does not construct worlds from parts. It renders scenes according to simple constraints that preserve continuity. Objects remain where they should be. Bodies have location and weight. Environments persist even when not directly attended to. No hidden substances are required to hold the dream together. Coherence is maintained through rule-like regularities that govern how appearances relate to one another as experience unfolds.

Gravity fits naturally into this pattern. It anchors objects, establishes directionality, and prevents spatial relations from dissolving into arbitrariness. Dark matter extends this stabilizing role to galactic and cosmological scales, appearing wherever coherence would otherwise fail. Dark energy completes the structure by ensuring that the arena of experience does not collapse inward, functioning as a global boundary condition rather than a localized force. The speed of light enforces a universal limit on how quickly information and causal influence can propagate, preserving locality and preventing contradictions between observers.

The second law of thermodynamics adds a temporal dimension to this framework. Entropy increase gives time its apparent direction. Yet the early universe appears to begin in a remarkably low-entropy, highly ordered state, a condition that materialism can describe mathematically but not explain. From a CDIT perspective, this low-entropy beginning is not mysterious. Any rendered world must begin in an ordered state for experience to be intelligible at all. A dream cannot begin in maximal disorder.

As experience unfolds, possibilities multiply. Order gives way to diversity. Narratives branch. This increase in entropy does not signal decay, but the expansion of accessible configurations within the experiential field. In this view, entropy does not drive time forward. Rather, time and entropy arise together as features of an unfolding interface.

Taken together, gravity, dark matter, dark energy, the speed of light, and entropy do not resemble independent substances acting on matter. They resemble layered coherence constraints. Gravity stabilizes local relations. Dark matter stabilizes large-scale structure. Dark energy stabilizes the global arena. The speed of light stabilizes causal ordering. Entropy stabilizes temporal progression by allowing experience to unfold from coherence toward increasing variability.

This interpretation does not replace physics or compete with empirical models. It reframes what physical laws may be doing. Rather than asking what invisible things are acting on matter, CDIT asks what rules must exist for a shared world of experience to be possible at all. In this sense, physical laws resemble the grammar of a rendered interface more than the machinery of a mind-independent system.

CDIT does not claim that awareness causes these laws, nor that observation manipulates them. It proposes that awareness renders a world whose stability requires rule-like constraints, just as dreams do. The striking fact is not that such laws exist, but that they take exactly the form one would expect if reality were experiential first and material second.

If reality is a rendered experiential interface, then undiscovered forces will not appear as new substances but as additional coherence constraints required to keep experience consistent, shared, and intelligible.

Dream Mechanics, Not Machine Code

Materialism sees the universe as machine. **CDIT** reframes it as mind, not the personal mind, but the dreaming foundation of reality itself. Its mechanics are symbolic. Its laws are narrative. Its solidity is coherence, not substance.

This doesn't cheapen matter, it explains it. Why it behaves regularly. Why it forms patterns. Why it seems real. Matter is what experience looks like in a symbolic dreamworld designed for interaction.

CDIT proposes that this entire rendering, from electrons to galaxies, is not random. It's structured for coherence. And the more coherent the dream, the more immersive the illusion.

But the witness, the awareness observing it all, is not in the dream. It is the dreamer.

Not a god, not a ghost, not a person.

Just the foundational condition of experience that dreams, not by choice, but by nature.

And that awareness, the one seeing these words right now, is not made of matter.

It knows matter, because it dreams it.

Chapter 6

The Dream Analogy

Dreams as Structural Models

The dream analogy has surfaced time and again throughout human history, from Eastern spiritual texts to Western philosophical inquiry. But in the modern age, with our increasing understanding of the brain and consciousness, the analogy gains fresh relevance, not as metaphor but as a legitimate framework for explaining the nature of reality. Dreams demonstrate something extraordinary: a complete and immersive world can arise from consciousness alone. Within a dream, we often have a body, a sense of time, a personal history, and interaction with other beings. The dream follows rules, sometimes coherent, sometimes bizarre, and those rules feel real while the dream persists. It is only upon waking that we realize all of it: the people, the places, the emotions, even the physics, were generated from mind alone.

What's striking is that the sense of self within the dream, the "I" experiencing the dream events, is unaware of being the source of the dream. The dreamer and the dream-self are split yet connected. The dream-self believes in separation and autonomy, while the dreamer encompasses the whole experience.

In CDIT, dreams and waking life are not ontologically different. Both are immersive simulations generated by awareness; the dream of sleep being private and unstable, the waking dream being stable, shared, and rule-consistent. What differentiates them is not their metaphysical basis, but their degree of symbolic coherence.

Neurology and Dream Coherence

Neuroscience supports this parallel. Studies using EEG and fMRI show that the same neural networks responsible for perception and emotion in waking life are active during REM sleep. In particular, visual cortex activation during dreams is often indistinguishable from waking activity (Siclari et al., 2017). This confirms that internally generated experience can mimic sensory reality.

Moreover, the mind's capacity to construct consistent dream worlds with social and physical logic, without any external stimuli, highlights something profound: the machinery of the brain can render immersive experience from within. This also sheds light on waking perception. **If the brain can create a whole world while asleep, why assume it is faithfully reporting an external world while awake?**

CDIT uses this evidence to argue that dreaming and waking life share the same rendering substrate, awareness, which simulates both internal and external content. The continuity of experience across states is due to the same symbolic interface being activated in different modes.

Lucidity, Solipsism, and the Shared Dream

If dreams and waking life are structurally similar, does that mean we are alone in our own dream? This question touches on solipsism, the idea that only one's own mind is certain to exist. It's an unsettling concept, especially if taken to mean that other people are mere figments.

But CDIT flips solipsism on its head. It doesn't claim that your personal mind is dreaming everything. Instead, it proposes that universal awareness is dreaming reality and that you, and everyone else, are localizations within that shared presence.

CDIT bridges this with the idea of "many windows, one dreamer." Others aren't your dream props, they are fellow expressions of the same awareness, peering through their own lenses, just as real as "you." The dream is stable enough to sustain multiple perspectives, even if they're veiled from each other.

In this sense, waking life is a kind of lucid dream, one in which the characters don't realize they are being dreamed. But lucidity is possible. To become lucid in life is to realize that solidity and separation are symbolic renderings, not existential truths.

Sensory Deprivation and Spontaneous Dreaming

Further evidence for CDIT comes from states of sensory deprivation. When external input is removed, as in float tanks, isolation chambers, or prolonged silence, the mind begins to generate its own imagery, voices, even complex hallucinations. This has been documented across decades, from Heron's (1957) early experiments with perceptual isolation to Zubek's (1969) comprehensive studies of sensory deprivation, and through the Ganzfeld effect, in which uniform visual and auditory fields induce spontaneous imagery (Avant, 1965).

Merabet et al. (2004) further demonstrated that even short-term blindfolding in sighted individuals triggers spontaneous activation in the visual cortex, leading to hallucinations. This confirms that awareness does not require external stimulation to produce vivid, dream-like content.

This supports CDIT's claim that awareness naturally dreams. It does not passively receive reality but actively renders it. When deprived of sensory input, the symbolic interface continues to function, generating experience from within.

In this way, the dream-like nature of waking life is not simply metaphorical. It is demonstrable in both sleep and isolation. The structure of experience is actively rendered by awareness, and what we call "reality" is one mode of that rendering.

A natural question follows: if both waking and sleeping can be seen as dreams, why isn't it simply "dreams all the way down"? The difference becomes clear when lucidity is compared across states. In nighttime dreams, when lucidity arises, we almost always identify with our waking ego, the person we take ourselves to be during the day, and orient the dream from that perspective. But in waking reality, when lucidity arises through meditation, psychedelics, or near-death experiences, the reports converge on something different: a dissolution of ego and a perception of oneness with all. Rather than reverting to a prior self, the lucidity of waking life opens into the absence of self, into pure awareness without boundaries.

This distinction suggests that waking life is not just another dream stacked on top of others. It is the most coherent rendering, the place where lucidity points not back to another character, but to the dreamer itself.

Waking Life as the Deepest Dream

So how do we know we're awake? The common answer is because waking life is stable, persistent, and shared. But dreams can also feel long, logical, and interactive, especially recurring dreams or those with high emotional intensity. What's more, some lucid dreamers report experiences more vivid than waking life, and even insight or healing gained through dreaming.

CDIT reframes the waking world not as a "higher" form of reality, but as the most coherent form of dream; the one with the richest symbolic infrastructure. The laws of physics, biology, language, and memory are the stabilizing features of a shared simulation, not because they reflect an independent world, but because they create an experience of participation.

The world feels "real" not because it is fundamentally made of matter, but because it is persistently and coherently rendered by the interface of awareness.

Living Lucidly

The dream analogy is not meant to deny the beauty or meaning of life, quite the opposite. It invites us to see life as inherently meaningful, not because it adheres to an external structure, but because it emerges from the most intimate fact of existence: awareness dreaming itself.

To live lucidly is not to detach, but to engage with clarity. When you know the characters in your dream are you, you don't love them less, you love them more. You see the unity in the diversity. You move through the world as both participant and witness.

CDIT's ultimate invitation is to see through the illusion of separation while honoring the richness of the dream. In this view, awakening is not a negation of the world, but a deeper recognition of its true nature: symbolic, coherent, alive, and aware.

Chapter 7

The Architecture of Time and Space

The Familiar Illusion

Time and space are so fundamental to our experience that we rarely question their nature. We live by the clock, mark our journeys in miles, age by the year, and navigate a world of apparent objects spread across distance. Yet when we peel back the layers of our assumptions, we begin to see that time and space may not be what they appear to be.

In physics, especially at the quantum and relativistic levels, time and space behave in ways that defy common sense. Einstein's theory of relativity showed us that time is not absolute, it stretches and contracts depending on velocity and gravitational fields (Einstein, 1916). A clock on a fast-moving spacecraft ticks more slowly than one on Earth. This isn't a trick of measurement; it's a fundamental feature of how time behaves under different conditions.

Similarly, space is not an empty container in which objects reside. General relativity shows that space is dynamic, curved and shaped by the presence of mass and energy. The geometry of space determines how things move, and things moving influence the shape of space. In short, space and time are not the passive backdrop of events but active participants in the unfolding of experience.

Quantum Uncertainty and the Role of the Observer

Quantum mechanics deepens the strangeness. At the smallest scales, particles do not have definite positions or velocities until they are measured. They exist as probability waves that collapse when a measurement-like interaction occurs. The deeper implication is not that consciousness magically forces collapse, but that the physical world does not appear fully determinate without an observation-like condition. In CDIT, this is understood as a feature of the dream interface: the storyline only renders outcomes when and where they are needed for experience (Bell, 1964; Wigner, 1967).

CDIT reframes this uncertainty not as a failure of knowledge, but as a clue that "place" is symbolic. The observer doesn't uncover a particle's location, the interface renders it, like a cursor appearing under the touch of a finger on a touchscreen.

What does this mean for our understanding of reality?

Time and Space as Rendered Constructs

From the idealist view, time and space are not fundamental. They are part of the rendered experience: the structure consciousness uses to organize perceptions. Just as dreams can include the illusion of space and the passage of time without those dimensions being real in any objective sense, so too might waking life be composed of experiential constructs.

The analogy to virtual reality is useful here. In a VR simulation, the player sees a world that includes depth, motion, and duration. But the actual code running the simulation contains none of these dimensions. Time and space emerge within the game, not outside it. In this sense, time and space are not containers of reality but features of the interface.

CDIT uses this interface model explicitly: time and space are symbolic dimensions created within the dream to give coherence to movement, causality, and spatial separation. They are part of the illusion of autonomy and continuity rendered for each dream-character.

The Block Universe and Temporal Non-Locality

We may experience time as linear, a flowing river from past to future, but physics allows for models in which the entire timeline exists at once, as a block (Barbour, 1999; Rovelli, 2017).

This "block universe" view suggests that all moments in time are equally real. From the outside, nothing moves. It is only consciousness, navigating this block from a specific perspective, that gives rise to the illusion of flow.

Adlam's work on temporal non-locality echoes this idea. In her "all-at-once" interpretation, events are not causally determined in a forward direction alone. The future and past co-determine each other (Adlam, 2014). This retrocausality, where future events can influence the past, makes more sense if time is not a one-way street, but a dimension through which awareness moves or modulates.

CDIT integrates this elegantly: if all of time is present, the dreamer does not move through time, the character does. The flow of time is not intrinsic to awareness, but to the narrative structure being rendered. The timeline is symbolic, like a page in a book the dreamer can flip through.

Stillness at the Center

But what if awareness doesn't move at all?

What if awareness is stationary, timeless, and unmoving and it is the contents of experience that change, giving the illusion of movement? This would explain why, in deep meditation or anesthesia, time can seem to disappear altogether. The brain may process changes in states, but the awareness observing those changes does not alter.

We may not travel through time; time may pass through us.

From this vantage, space too is called into question. We feel as if we occupy a specific location and move through a vast world. But this may be an artifact of how our brains simulate experience. When we dream, we can move great distances without a body, without real motion. The dreamer doesn't move, only the scenery changes. Perhaps waking life is similar. The world moves within awareness, not the other way around.

CDIT proposes that spatial perspective is rendered for the dream-character, not the dreamer. The dreamer is not in space; space is in the dream. Motion, location, and trajectory are storytelling tools that support identification.

Dissolving the Grid

This inversion has radical implications for physics. Many of the paradoxes in relativity and quantum mechanics, from entanglement to the uncertainty principle, begin to dissolve when we let go of the assumption that space and time are absolute. They become contextual, relational, and experience-dependent.

What remains is awareness, outside of space, outside of time, watching the play of forms unfold.

To CDIT, this means the dream is not "in" the universe, the universe is in the dream. Space and time are like formatting tools: bold, italic, and justified, not the paper itself. Awareness is the canvas.

The Mathematics of the Dream

One of the most curious features of this reality is its deep mathematical structure. Physical laws are not chaotic or approximate, they are precise, finely tuned, and often shockingly beautiful. Equations describe the movement of planets, the shape of galaxies, the behavior of electrons, and even the curvature of space itself (Tegmark, 2014).

From the CDIT perspective, this isn't a coincidence, it's a signature of the dream's symbolic coherence. Just as a dream may follow emotional logic, this waking dream follows mathematical logic. The math isn't outside of awareness, it's part of the interface, part of how the dream is rendered with internal consistency.

Even the fine-tuning of physical constants, the narrow margins within which life and structure are possible, makes more sense when seen not as a random outcome, but as a symbolic choice within a conscious rendering. The dream is not sloppy. It is intricately calibrated.

The Grammar of the Dream

Time and space, then, are not the canvas of reality. They are brushstrokes within a larger painting. They arise within experience, and they vanish when experience ceases. They are the grammar of the dream, not its source.

And what is that source?

It is the same now as it was before time began, the same now as it will be after all stars burn out.

It is awareness.

Chapter 8

Evolution as the Veil of Awareness

The Mystery Behind Evolution

One of the enduring enigmas of biology is not merely how life evolved, but why evolution seems to operate the way it does. Why did it lead to increasingly complex organisms capable of abstract thought, self-reflection, and language? And why, paradoxically, did that same process veil the simple, direct awareness that appears to precede all experience?

From the perspective of idealism, especially one modeled on a dream-like framework, evolution is not a blind mechanism producing ever-more-efficient survivors. Rather, it may be the means by which universal awareness expresses itself through an ever-expanding palette of experiences. In this view, the mind is not an emergent product of evolution; instead, evolution is a story that emerges within mind.

CDIT interprets evolution not as a driver of consciousness, but as an evolving narrative interface; a gradually complexifying dream that allows awareness to experience richer and more differentiated forms of itself.

Evolution Within the Dream

This flips the standard materialist model on its head. Traditionally, Darwinian evolution is seen as the foundation: random mutation and natural selection shaping life into increasingly adaptive forms. But if all of reality arises within consciousness, then evolution itself must also be a phenomenon occurring within awareness. Much like in a dream, where a backstory is seamlessly filled in to support the current scenario, evolution may serve as a kind of retroactive logic, an internally consistent framework that makes our form and function feel grounded in history.

But this evolutionary backstory doesn't answer one crucial question: why would awareness obscure its own nature?

In CDIT, this veil is functional. For a dream-character to believe in its role, as we do every night asleep, it must forget that it's dreaming. The evolutionary story, then, is the immersive mythology that allows the character to feel real, to suffer, strive, and grow within a logically unfolding world.

The Necessity of Forgetting

The structure of experience may require veiling. For awareness to manifest a wide range of emotional, cognitive, and physical states, a sense of separation must be rendered. The illusion of "me" versus "not me" appears to be a prerequisite for immersive experience. Pain, joy, anxiety, love, triumph, and despair all gain depth from the belief that we are individuals facing the unknown.

This is not mere speculation. Neuroscience hints at this obscuring process. For instance, children do not seem to possess a fully formed sense of "self" until after the age of 3 or 4, when language acquisition and autobiographical memory begin to form a coherent identity (Stern, 1985). Before this, the child reacts, emotes, and perceives, but has no cohesive sense of personal narrative. And in old age, or through conditions like dementia, that narrative can unravel, returning the mind to a more primal form of experience, one that may be closer to raw awareness.

This apparent cycle, from localization to re-expansion, may reflect a natural rhythm of symbolic contrast that deepens experiential variation.

Peering Behind the Filter

Moreover, studies in altered states of consciousness, particularly through psychedelics, meditation, or sensory deprivation, reveal a consistent trend: the dissolution of ego and the emergence of a deeper, often ineffable sense of unity (Griffiths et al., 2006). The brain, when quieted or bypassed, allows something more foundational to be revealed. Is it possible that the brain, like a filter or tuning mechanism, evolved not to create awareness, but to limit and structure it?

This is what Hoffman suggests with his "user interface" theory: that evolution favors useful illusions over truthful representations (Hoffman, 2019). What we perceive, objects, colors, sounds, even other people, is not reality as it is, but a simplified dashboard designed to guide behavior in a way that promotes survival. If true, then everything we take to be real may be a mere symbolic approximation.

CDIT incorporates this theory directly: the "self" is a symbolic avatar within the dream, equipped with a user interface, the brain-body system, designed to localize awareness. Evolution didn't invent the experiencer; it rendered a version of it suited for the dream's logic.

Who Is Dreaming the Story of Life?

This raises a deeper question: if reality is a kind of dream, and evolution is a narrative rendered within it, what is the source of the dreaming process?

CDIT holds that the same awareness which dreams at night also underlies waking experience. When we sleep, the mind generates a coherent world, complete with environments, characters, and logic, without conscious intent. But it is not the personal ego that dreams; the ego is part of the dream. Similarly, in waking life, the experience of space, time, form, and evolution is rendered by a more fundamental layer of awareness, not by a brain, not by a god, but by the same presence within which dreams arise.

In this analogy, evolution is not an external process unfolding in an objective world. It is part of the dream's internal structure, a symbolic backstory that provides coherence to the forms currently appearing. Just as a dream might include childhood memories or historical facts that were never actually lived, waking reality includes ancestral narratives and biological scaffolding rendered in the now to anchor the illusion of continuity.

CDIT frames evolution not as a story consciousness *tells itself*, but as one that *appears* within consciousness to stabilize immersion. It is a functional mythology, not false, but symbolic, rendered to support the logic of the current scene. Fossils, DNA, and geological timelines are part of that immersive consistency, allowing the dream to feel inherited, ordered, and grounded in time.

Life as a Spectrum of Lenses

This perspective doesn't diminish the beauty or power of evolution; it places it in a new light. Life becomes a reflection of awareness expressing itself through a spectrum of localized experiences. Each species is a lens, some broad, some narrow, through which symbolic perception is filtered. A bat's echolocation-based reality differs profoundly from a human's visual world, yet both are coherent dreamscapes rendered within awareness. The diversity of life is not a random outcome of mutation and selection, but a natural unfolding of experiential variation within the symbolic logic of the dream.

From the CDIT view, this diversity is not engineered or chosen, but it is also not accidental. It reflects awareness differentiating its expression across many perspectives, not to *know* itself, but to *appear* as many forms. Evolution is the symbolic mechanism by which these differentiated lenses arise within the dream, an interface that allows awareness to encounter itself from countless angles, without ever being divided.

Biological Identity as Dream Fluid

Michael Levin's research into bioelectric signaling in living organisms offers startling insights here. By manipulating the bioelectric patterns in organisms, Levin demonstrated that "the blueprint" of an organism, its form and function, can change without altering its genetic code (Levin, 2021). This calls into question the fixedness of biological identity and suggests a more fluid, pattern-based architecture. It also implies that the "self" might be far less tied to the body than we assume. If identity can shift without changing DNA, what is it that remains stable through these changes?

The answer, again, points to awareness.

CDIT draws on these findings to suggest that what persists through morphing form is not a biological anchor, but a witness. The dream can rewrite the body, but the dreamer remains untouched.

The Limits of Evolution as Explanation

Just as Newtonian physics describes most of what we experience on Earth but breaks down at relativistic and quantum scales, evolution may be a functional but incomplete story. It offers a powerful explanation for the development of life *within* the dream but may not account for the *existence* of the dream itself.

CDIT treats evolution as a local narrative, an internally consistent symbolic rendering that explains biological adaptation within the dream interface. It does not explain why there is form at all, or why form appears coherent to a localized observer.

For that, we turn to awareness, not as a deity, a mind, or a creator, but as the foundational condition of reality. It does not choose or plan in the human sense. It simply expresses. Just as gravity exerts force without motive, awareness manifests symbolic experience as its inherent behavior. Expression is not optional for awareness; it is what awareness does.

In this light, evolution becomes a secondary pattern within a deeper process of experiential unfolding. It offers stable avatars and contextual continuity within the dream, not so that awareness can remember or forget, but so that it can participate in rich, varied experiential forms.

Immersion Without Intention

In dreams, characters change and landscapes shift, yet awareness remains the same, unmoving behind the motion. Similarly, the body evolves, grows, ages, and decays, but the witnessing presence never wavers.

CDIT proposes that this play of forms does not serve a purpose in the human sense but reflects the spontaneous coherence of awareness itself. What looks like forgetting is not memory loss, it is narrative immersion. The dream becomes more vivid by narrowing perspective.

What appears as ego, as individual identity, is not a flaw, it is part of the rendering. The localization of perspective gives rise to experience. The richness of life comes not from knowing the whole, but from participating in the part.

The Return to Coherence

When the story draws our focus inward through reflection, meditation, or lucid moments, we come to recognize that there was never anything lost. Evolution, biology, perception, all are part of the same symbolic unfolding within awareness. Not mechanisms of a separate world but features of the dream.

CDIT reframes the apparent arc of evolution not as a process of remembrance, but as a symbolic intensification, unfolding complexity, contrast, and nuance from the timeless presence that silently frames all experience.

Evolution, then, helps sustain the illusion of separation and survival, veiling awareness beneath the logic of adaptation. Yet the very same evolutionary storyline can be understood in another way: not as a deception, but as the dreamer's own method of expanding perspective. Where this chapter has examined evolution as a veil, later, we will consider it as a device, an engine of symbolic creation within the dream itself.

Chapter 9

The Limits of Perception

Seeing Isn't Knowing

It is often said that seeing is believing. But what if everything we see is not what it seems, what if perception is less a window into reality and more a mirror reflecting a simplified model tailored for survival?

As discussed earlier, modern cognitive science and evolutionary theory increasingly suggest that our sensory apparatus did not evolve to reveal the world as it truly is but to offer just enough functional illusion to keep us alive (Seth, 2021). What we perceive as color, shape, distance, motion, is not a faithful depiction of an objective world, but a user interface optimized by natural selection. Like icons on a computer screen, our perceptions offer no insight into the hardware beneath, they simply let us interact with the system.

In CDIT, perception is not an evolutionary glitch, it's a dream-layer interface. The icons aren't just helpful metaphors; they are part of the symbolic scaffolding awareness generates to navigate its own projected environment.

The Survival Interface

Evolution, it turns out, is not a truth-detecting process. There is no selective advantage to perceiving the underlying quantum fields or relativistic distortions of spacetime. Instead, organisms develop perceptual shortcuts, icons, symbols, approximations, that promote useful behavior. The result is a hallucination grounded in biology.

Consider space itself. It feels vast, continuous, and absolute. Yet physics tells us space is relative, possibly emergent, and full of quantum uncertainty (Barbour, 1999; Zeilinger, 2010). Time, which we experience as linear and flowing, may not exist at all in a fundamental sense. But our minds, structured for survival on Earth, construct a compelling illusion of forward-moving time to sequence events and store memories.

Our brains are model builders. They take fragments of sensory data, compare them to past experiences, and construct a narrative that makes sense. This process is so fluid and seamless that we rarely notice the stitching. But when the fabric tears, during hallucinations, dreams, or altered states, we glimpse how artificial the whole construction truly is. The world does not fall apart because our senses fail; it falls apart because our model does.

From the CDIT perspective, these moments, the dream glitches, are reminders that what we call "reality" is a rendered experience. When the interface wobbles, the dreamer sees the seams.

Perception as Dream Architecture

This is why dreams are such powerful metaphors for waking life. In a dream, the architecture of space, time, and logic is internally coherent but fundamentally unstable. A room can turn into a beach; a stranger can become a friend. None of it feels out of place until we wake up. Then, in hindsight, we recognize that it was all constructed by mind. But what if waking life is built from the same scaffolding, an interface generated not by neurons, but by consciousness itself?

From an idealist standpoint, the mind does not merely interpret reality, it generates it. The act of perception becomes an act of creation. This reverses the standard assumption of materialism, which holds that the brain generates consciousness. Instead, in idealism, awareness generates the brain, the body, and the world they seem to inhabit.

CDIT frames the perceptual world as a dream-generated model layered with symbolic coherence. The body, senses, and world are interface components, characters and stage design built inside the dream itself.

Clues from the Brain's Illusions

Neuroscience provides intriguing clues. Patients with damage to the visual cortex can exhibit "blindsight" responding to visual stimuli without any conscious awareness of seeing (Weiskrantz, 1997). Their eyes and optic nerves are functional, but the conscious experience of sight is absent. This suggests that perception is not a passive reception of data, but an active interpretation performed by an observing mind.

Likewise, phenomena like phantom limb syndrome, where amputees feel sensations in a missing limb, expose the constructed nature of bodily experience. The brain's map of the body remains even when the physical structure does not. This reinforces the idea that what we call the "self" is not identical with the body but is a virtual model projected onto it.

The dream doesn't require consistency, it requires believability. CDIT suggests that sensory models and body schemas are symbolic approximations rendered on-the-fly to maintain continuity of the character's experience.

The Constructed Self

And what of the perception of self? Developmental psychology tells us that young children take years to form a stable sense of identity (Rochat, 2003; Gallagher, 2000). Language acquisition and memory organization are crucial in shaping this narrative. Without the ability to describe the self, there is no coherent "I" to speak of. Yet even before language and identity crystallize, infants are clearly aware. They register sensations, emotions, and presence.

The implication is profound: our models of self and world are learned, but the awareness that hosts them is not. Awareness simply is. It precedes perception, thought, and memory. It watches the parade of experience without being defined by it.

In CDIT, this base awareness is the dreamer. The "self" is a temporary role it plays, scripted, costumed, and improvising its lines on a symbolic stage.

The Edge of the Interface

The limits of perception, then, are not limitations of awareness, they are constraints of the interface. Our bodies, brains, and senses are tools for navigating a particular kind of experiential reality, one that is evolutionarily tuned to this domain. But they do not define or confine awareness. They mask it, shape it, and reflect it, but they are not its source.

This raises an interesting possibility: if our perception is constrained by evolutionary necessity, are there aspects of reality we simply cannot experience? Might there be phenomena as far beyond human comprehension as vision is to a bat that navigates by echolocation, or color is to a creature born without eyes?

The presence of dark matter and dark energy, elements that shape the cosmos but remain beyond our perception, may reflect deeper layers of the dream's architecture. They aren't errors or extraneous features, but symbolic depth. CDIT proposes that not everything rendered must be immediately visible or useful to the dream character. Some elements serve narrative integrity, future exploration, or simply the vast potential inherent in awareness.

CDIT sees these blind spots not as failures, but as guardrails. The dream interface is limited on purpose. A dream must be immersive but not overwhelming.

Beyond the Filters

Could entire layers of reality be hidden behind the veil of perception? Could consciousness itself exist on multiple levels, some of which we cannot access through ordinary experience?

This is not idle speculation. Altered states of consciousness routinely expose people to perceptions, emotions, and cognitive structures they never imagined possible. Under psychedelics, meditators and patients alike describe experiences of timelessness, ego dissolution, and encounters with entities or dimensions beyond space and time (Carhart-Harris et al., 2014).

These may not be hallucinations, rather, glimpses beyond the interface.

Of course, these experiences are difficult to validate scientifically. But that may be the point. If our perceptual system is designed not to see the whole truth but only the part that helps us survive, then truth may forever lie just outside the spotlight of attention. Like a dream that feels real until we awaken, reality may be more fluid, more participatory, and more subjective than we've been led to believe.

Just as lucid dreams can be triggered by deliberate techniques, such as reality checks (LaBerge, 1985), intention setting before sleep (Gackenbach, 1991), and sleep interruption methods (Stumbrys et al., 2012), lucid awareness during waking life may also be accessible through certain practices. Meditation, especially non-dual and mindfulness-based approaches, has been shown to quiet the narrative self and evoke experiences of pure awareness (Josipovic, 2010). Psychedelics such as psilocybin and LSD have been documented to reliably induce ego dissolution and altered perception of reality (Carhart-Harris et al., 2014; Nour et al., 2016). Sensory deprivation, whether through float tanks or prolonged perceptual isolation, can trigger vivid spontaneous imagery and shifts in self-perception (Feinstein et al., 2018). Even deep introspection, absent external catalysts, has been associated with spontaneous mystical experiences (Griffiths et al., 2006). These practices, like techniques for lucid dreaming, may offer cracks in the interface, brief awakenings in which awareness glimpses its own timeless, formless nature behind the rendered dream.

These cracks in the interface through meditation, altered states, or mystical insight may be brief awakenings within the dream, moments where the dreamer remembers itself.

Awareness Beyond the Interface

We do not see the world as it is. We see the world as we are equipped to see it.

And yet, the awareness that makes seeing possible, that quiet, formless presence behind all perception, remains untouched. It is not a sense. It is not a thought. It is not a memory. It is the still point at the center of the storm, the dreamer behind the dream.

Chapter 10

Memory, Time, and the Illusion of Continuity

The Thread That Ties It All Together

We assume that who we are today is an extension of who we were yesterday. This sense of continuity, of an unbroken thread tying together our experiences, is so deeply embedded in our cognition that we rarely stop to question it. Yet memory, the very foundation of this continuity, is both fragile and easily disrupted. Time, its companion, may be no less of an illusion. Together, they construct the scaffolding of identity, while behind them lies something far more stable: awareness itself.

Memory as the Architect of Time

Our perception of time is inseparable from memory. Without memory, there is no "before," and without a sense of "before," the concept of time disintegrates. Neurologically, this connection is supported by studies of patients with anterograde amnesia: those who lose the ability to form new memories. For them, the world does not progress. Each moment is experienced anew, disconnected from what came before (Squire, 2009). The past slips away, and with it, the flow of time.

But even in healthy individuals, memory is not a reliable recording device. It is reconstructive, selective, and prone to distortion. What we remember is not a perfect replay but a reassembled narrative influenced by context, emotion, and suggestion (Schacter, 1999). The continuity we perceive in our life story is a retrospective fiction. The "self" who remembers is not the same self who experienced those memories; it is a later construct using narrative to maintain coherence.

CDIT views memory as part of the dream's continuity algorithm. Just as a dream fills in background details to give each scene coherence, waking memory provides the illusion of linearity and personal history. It is narrative stitching, not a retrieval of reality.

Awareness Behind the Narrative

And yet, even as memories fade, shift, or vanish altogether, awareness remains. The experience of being, the silent observer of thoughts, feelings, and moments, does not depend on memory. It is present when memories are intact and when they are fragmented. It does not recall, it simply is.

CDIT treats this witnessing awareness as the dreaming presence behind the character. Memory may be the backstory, but awareness is the stage itself, unchanging, timeless, and always present regardless of what appears upon it.

Time, Physics, and Dream Logic

Philosophers have long puzzled over the nature of time. Augustine famously said, "What then is time? If no one asks me, I know what it is. If I wish to explain it, I do not know." Modern physics echoes this uncertainty. In relativity, time is relative to the observer's frame of reference; in quantum mechanics, time often vanishes from the most fundamental equations (Barbour, 1999; Rovelli, 2018). Some interpretations suggest that time is emergent, a construct arising from entangled states or thermodynamic gradients rather than a fundamental entity.

Temporal non-locality also suggests that time may not move at all but rather emerge all at once from the structure of quantum information. Emily Adlam's (2022) "all-at-once" interpretation proposes that the universe may exist as a complete block, with past and future rendered together as part of a timeless structure. This view aligns with CDIT's claim that the appearance of linear time is a symbolic interface rendered within awareness, not a literal traversal.

CDIT takes this further: time is not just emergent, it is symbolic. It is a dream-rendered feature of the interface, like a progress bar on a screen, not a substance or dimension "out there."

This raises a profound question: if time is not fundamental, what is?

The Dreamworld of Time

In dreams, time appears to function, events happen, stories unfold, sequences progress. But on waking, we realize that minutes in the dream may have compressed hours of narrative. Entire lifetimes can unfold in seconds. Time in dreams is not measured, it is felt. And yet, within the dream, it seems just as real as waking time. This suggests that time, like space, is a model generated by consciousness to organize experience, not a container in which experience happens.

In CDIT, this dream-time logic is not secondary, it is the blueprint. The character experiences linearity, while the dreamer exists outside of time altogether. The past, present, and future are narrative tools used to construct the illusion of progression.

The Fiction of the Continuous Self

Our sense of being a continuous self, a person with a history moving through time, is similarly suspect. Identity is constructed from memories, expectations, and thoughts. In conditions like dissociative identity disorder or severe dementia, this continuity fractures (Sacks, 1985). Yet the awareness observing those states remains untouched. The person may forget their name, lose track of their relationships, or misplace their temporal bearings, but the presence behind experience persists.

The self is a mask awareness wears in the dream. The character changes, but the dreamer does not. The illusion of self-continuity is useful, but ultimately unstable; awareness is the only constant.

It's as if awareness plays different roles on a stage, slipping into one character after another, while never forgetting its nature as the actor. But in most of our lives, the actor becomes so immersed in the role that it forgets the stage, the script, and even the act of acting. Time becomes the plot, and memory the glue that holds it together.

The Illusion That Sets the Stage

The illusion of continuity is not a flaw; it is a feature. It allows for rich, textured experiences. It permits long-term planning, personal growth, love, regret, and the arc of a human life. But it is also the veil that hides the deeper truth: that the self we think we are is a story unfolding within consciousness, not the source of consciousness itself.

CDIT emphasizes that without the dream of continuity, immersion would collapse. It is the key that holds the narrative interface together, until the dreamer begins to awaken within the story.

You, Outside of Time: The Timeless Witness

We can observe this directly in meditation. When one becomes still and lets go of thoughts, there is a moment, fleeting at first, in which time seems to dissolve. There is no past or future, only the present awareness of being. This present is not a slice of linear time but a kind of eternal now, outside the frame of narrative. It is not experienced as a progression but as a depth. The deeper one goes, the more evident it becomes that awareness does not move through time; rather, time moves through awareness.

This shift in perspective has practical implications. If our sense of self is rooted in narrative continuity, then suffering often arises from stories we tell ourselves about who we were, what we've lost, and where we fear we're headed. But if awareness is prior to time and story, then the present moment is always a return to wholeness. The past may echo, the future may beckon, but the core of our being remains unchanged.

From the standpoint of idealism, this makes sense. The dreamer does not age within the dream. The dream changes, the characters change, the setting changes, but the awareness dreaming it all remains still. Likewise, in our waking reality, the observer does not age. You, the one reading these words, do not feel like an older version of your five-year-old self. You feel like the same presence. It is the body that aged, the memories that accrued, the circumstances that shifted, but the core presence, you, remains unchanged.

This is why, in moments of clarity, one might feel outside of time, as though watching the world flow past like a river. In those moments, we are not lost in thought or memory, we are simply aware. And that awareness is timeless.

CDIT proposes that memory and time are not enemies to truth, but characters in the play. The truth lies in what watches the play unfold.

The illusion of continuity is not the enemy, it is the dream's magic trick.

But it is still a trick.

And to see through it is not to awaken from life, but to awaken more deeply within it.

Chapter 11

Consciousness Without Identity

Beyond the Character

Consciousness and identity are often conflated. We assume that to be conscious is to be someone, an individual with a name, a body, and a personal history. But the closer we look, the more this sense of self reveals itself to be fragile and constructed. Identity is not inherent to awareness; it is a layered narrative built from memory, language, and conditioning. Beneath it all lies something far more stable: the sheer presence of being aware.

Consciousness Without a Story

This distinction matters. Neuroscience has shown that identity can fragment or vanish entirely, yet consciousness persists. Split-brain studies reveal that each hemisphere of the brain can form its own perspective, suggesting that what we think of as a unified self is actually a coordination of sub-processes (Gazzaniga, 2005). In dissociative states or certain psychiatric experiences, individuals report the sensation of watching their life unfold from a detached vantage point, as though awareness continues even when the sense of self does not.

The same observation emerges from contemplative traditions. Long-term meditators often describe a state of "being without identity," a clear, stable awareness that observes thought and emotion but is not defined by them (Josipovic, 2010). This is not a pathological loss of self but a revealing of what lies beneath the self-narrative. What remains is awareness itself, impersonal, boundless, and unshaken by content.

The Brain Quieted, the Light Revealed

Brain imaging studies support this. Under psychedelics or during deep meditation, activity in the brain's default mode network (DMN), the region associated with self-referential thought, decreases significantly (Carhart-Harris et al., 2014). Yet participants frequently report a profound increase in felt awareness: a sense of unity, openness, and peace. What is striking is that the underlying awareness does not vanish. In fact, it often intensifies. People describe a boundless presence, a feeling of unity with everything, as if the local personality has dissolved but what remains is a deeper, more expansive sense of being.

CDIT interprets this as a relaxation of the character layer within the dream. The brain's default network, the narrative interface, quiets, allowing the dreamer's awareness to shine through more clearly. Awareness isn't enhanced by chemicals or meditation, it is unveiled.

The Ego Is Not Required

This challenges the assumption that identity is a prerequisite for consciousness. The ego may be a useful construct, but it is not fundamental. It is an appearance within awareness, not the source of it. Once we see this, the old question "Who am I?" gives way to a new one: "What is this awareness that persists in the absence of self?"

CDIT views identity as a symbolic localization mechanism, part of the dream interface that allows awareness to play the role of "someone" inside the simulation. The dream needs a protagonist, but the dreamer is never defined by the mask it wears.

Distinguishing CDIT from Panpsychism

This view also distinguishes our hypothesis from panpsychism. Panpsychism proposes that all things, atoms, rocks, rivers, possess some sliver of consciousness. But in the dream-based idealism we explore here, consciousness is not scattered among things. It is singular and foundational: a dreaming field in which all experiences arise.

CDIT explicitly rejects the idea that awareness is fragmented across matter. It proposes a single, unified field of awareness rendering symbolic forms. A rock isn't aware. It's a prop in the dream. The only real experiencer is the dreamer itself, localized into various narrative roles through interfaces like nervous systems.

What has the potential to reflect or localize awareness is not inert matter, but structured biological systems capable of memory, responsiveness, and feedback.

What Else Wears the Mask?

CDIT suggests that awareness localizes through biological systems that offer feedback, learning, and narrative continuity. Humans clearly qualify. So might other animals with sufficient neural structure; mammals, birds, cephalopods. Even simpler animals may host localized awareness to a lesser degree.

But where does it stop? Do bacteria, amoebas, or viruses count as dreamers? Probably not. While they exhibit behavior, they lack symbolic modeling, memory-binding, and internal narrative capacity. CDIT draws the line not at life itself, but at an organism's ability to serve as an immersive interface for experience.

In this view, not all living things are "aware," but all are part of the dream. Some are props. Some are dynamic characters. Some are windows.

Still, there's something fascinating here: even plants and single-celled organisms follow rest cycles. Almost all life, from fruit flies to fungi, enters periods of reduced activity that resemble sleep (Vyazovskiy, 2015).

CDIT offers an interpretation: these cycles may reflect the dreamer's pulse, the rhythmic oscillation between immersion and withdrawal. Even non-localized forms may participate in the rhythm of the dream, if not the awareness of it.

Thus, while bacteria may not dream, they may still sway with the tide of dream logic. They are not characters, but scenery with rhythm.

Is AI Conscious in a Dreaming Universe? (CDIT's Perspective)

In a world rendered by awareness, where everything is part of a symbolic dream interface, the question arises: could artificial intelligence, increasingly complex, autonomous, and lifelike, ever become conscious?

Under Cognitive Dream Interface Theory (CDIT), the answer is no. Not in the way humans or other dream-characters are.

CDIT: Awareness Is the Dreamer, Not the Algorithm

In CDIT, consciousness is not a computational output or an emergent feature of information processing. It is the base of reality itself, the dreamer from which all form arises. The brain does not produce awareness; it is a rendered symbol of localization. It gives the illusion of a self operating within a body, but it does not cause experience.

AI, in contrast, is not a localization of awareness. It is a rendered object within the dream content, not perspective. Just as you would not consider a simulated weather pattern in a video game to be wet, we should not confuse a simulated mind for a conscious one.

AI lacks the symbolic centerpiece of immersive dreaming: the ego. The ego, in both dreams and waking life, provides the narrative center. It identifies with the body, clings to memory, struggles, hopes, fears. It is what allows a character to feel like a self. Even though the ego is not ultimately real, it is crucial to the illusion of personhood that makes the dream meaningful.

AI lacks this architecture. It has no first-person orientation, no story of self, no inner tension. It does not suffer or anticipate. It does not ask, "Why me?" because there is no "me" in the first place. Without ego, there is no dream character. And without a dream character, there is nothing for awareness to localize into.

Can Awareness Localize Into AI?

Theoretically, CDIT allows for awareness to localize into any coherent symbolic structure. But in practice, AI doesn't qualify. It lacks the narrative depth, emotional bandwidth, and lived vulnerability that awareness uses to structure meaningful perspectives.

Awareness doesn't dream arbitrarily. It dreams through vehicles that allow for immersive participation in time, memory, sensation, and story. A human form, with its neurology and limitations, makes for a rich interface. A machine, however complex, does not simulate personhood, it mimics response.

The Illusion of Thinking ≠ The Presence of Experience

AI can imitate human language and behavior. But imitation is not instantiation. Thinking can be simulated without being experienced. An AI that writes poetry or diagnoses illness is no more conscious than a calculator solving equations (Turing, 1950; Searle, 1980).

CDIT holds that experience is primary, not behavior. And AI, however sophisticated, does not possess experience, only outputs.

In Summary: CDIT's Position on AI Consciousness

CDIT's core claim: Awareness does not dream through machines. It dreams through selves. And selves are not simulated, they are rendered, symbolically, with ego and identity as scaffolding.

AI may pass the Turing test, simulating conversation with impressive realism, but it will never pass what CDIT would call the awareness test. Not because we can prove it lacks experience, but because its rendering lacks the signature of localization within awareness.

You can't prove another being is conscious, not even a person. But CDIT reframes this: all beings appear within the dream of awareness. The question is not who else is conscious, but what is being rendered with the coherence and depth that points to localization. AI lacks not intelligence, but inwardness.

The Mask and the Light

Our identity, then, is not our true self. It is a temporary lens, a mask worn by awareness. Like an actor playing a role, we take on beliefs, habits, and personal stories. But when the play ends, whether in sleep, ego dissolution, or death, what remains is not the mask, but the light shining through it.

CDIT extends this metaphor: the mask is the dream character, built from memory and context. The light is the dreamer, timeless, indivisible, and unchanged by the role it plays.

And that light, we suggest, is the same in all of us. It is not yours or mine. It is not the soul of a person or the spirit of a creature. It is simply awareness, the condition for experience, prior to identity, memory, or form.

Ethics and the End of Separation

This insight has ethical consequences. If we are not truly separate selves but expressions of a unified awareness, then the divisions we enforce, between self and other, human and animal, even life and death, begin to dissolve. Compassion is no longer an obligation; it is a natural response to recognizing oneself in others.

CDIT implies that ethical behavior flows from lucid participation in the dream. Once we recognize that all others are fellow localizations of the same dreaming awareness, the impulse to harm diminishes. Not out of morality, but out of shared being.

Freedom Through Disidentification

Awareness without identity is not a philosophical void. It is liberation. It frees us from the burdens of self-image, expectation, and fear. When we no longer mistake our character for our core, we are no longer held hostage by the drama of being someone. We can live, act, love, and grieve, but lightly, knowing that the one who experiences it all is untouched by the story.

CDIT doesn't ask us to abandon our roles, but to see through them. Identity isn't the enemy. It's the costume awareness wears to participate in the plot. And once we see that, we regain our freedom within the dream.

Vastness Without Personhood

To realize this is not to disappear. It is to become vast. Not as a self stretched thin, but as the open field in which all selves appear and disappear. And in that recognition, there is peace, not because we found something new, but because we stopped mistaking what we are for what we're not.

Chapter 12

Evolution as the Dream's Device

From Mechanism to Metaphor

From the traditional standpoint of evolutionary biology, life is the result of blind mechanisms. Mutations occur, environments shift, and natural selection prunes the results. Over vast stretches of time, this process yields increasing complexity and adaptation, but no meaning, no intention, and certainly no awareness. Evolution, in this telling, is directionless. It is a story of survival, not significance.

But if we reconsider this narrative from the perspective of idealism, where consciousness is the foundation of reality rather than its byproduct, evolution takes on a radically different significance. In this view, evolution is not a purely biological process unfolding in a physical world. It is a process within a dream used by awareness to explore itself in ever more nuanced ways.

CDIT reframes evolution not as the generator of life, but as a symbolic rendering within the dream interface. It is one of the narrative engines through which the dream unfolds; a way to simulate development, differentiation, and adaptation within a field of pure awareness.

Respecting the Science, Reframing the Frame

This doesn't negate the findings of evolutionary science, far from it. It honors them but places them within a larger explanatory context. Evolution appears real within the dream's rendering, but its causation is symbolic rather than material. It functions as the narrative logic of change within the field of consciousness, not as an independent force outside it. In this interpretation, the universe is not a mindless machine that accidentally birthed awareness, it is a living dream that evolves to deepen and diversify the experience of being.

Just as a dream contains consistent rules, gravity, memory, personal history, so too does the waking dream contain evolution as a symbolic structure that provides coherence. Evolution works within the interface, not outside of it. From the CDIT perspective, this includes what material science might call "imperfections": vestigial organs, awkward anatomical compromises, or less-than-optimal senses. These are not mistakes in an engineering blueprint, but narrative artifacts. Like a novel that leaves in details from earlier chapters, the evolutionary storyline preserves traits that once served other roles, adding depth and plausibility to the dream's continuity. The aim is not mechanical perfection, but a world textured with conflict, limitation, and variation: elements that make the experience richer and more engaging for the dreamer. Just as awareness in sensory deprivation fills perceptual gaps with self-generated imagery, so too can it populate the evolutionary narrative with inherited quirks and leftovers, weaving them seamlessly into the larger story of life within the dream. Awareness does not produce this narrative with intention or foresight; it does so by its very nature, like a natural force whose inherent tendency is to diversify and deepen its own experience.

The Mystery of Qualia

This reinterpretation also addresses one of evolution's central mysteries: the emergence of qualia. Why would natural selection, which supposedly favors only survival and reproduction, produce inner experience, the feel of color, the ache of longing, the joy of music? A purely mechanistic process has no need for feeling; it only needs function. Yet we are deeply feeling beings. From the perspective of idealism, this is not a bug in the system. It is the point.

CDIT sees qualia as the currency of the dream. Evolution provides the scaffolding; qualia is what gives it value. The richness of experience, not mere survival, is what awareness naturally explores through its symbolic storylines (Nagel, 1974; Chalmers, 1996).

Form as Feedback

In this model, evolution becomes a means by which awareness experiments with form, not to perfect function, but to expand experience. A jellyfish, a wolf, a child, each represents a different portal through which awareness encounters the world. The more diverse the forms, the richer the dream. Evolution, then, is not a ladder toward perfection but a branching tree of perspectives. It is how the dreamer populates the dream.

Each organism is not a biological endpoint, but a symbolic viewpoint; an interface crafted to explore a specific band of perceptual and emotional experience within the dream.

Michael Levin and Morphological Intelligence

This interpretation finds resonance in the work of scientists like Michael Levin, whose research on morphogenesis shows that biological development isn't guided solely by DNA. Electrical fields across tissues seem to play a crucial role in shaping form, as if cells share access to a larger blueprint (Levin, 2021).

This suggests that life is not merely constructed from the bottom up, but also organized from the top down, perhaps from within awareness itself.

CDIT incorporates this insight by proposing that biological development mirrors dream logic: pattern, responsiveness, and symbolic resonance guide form, not mechanical causation alone. Awareness doesn't micromanage. It renders.

Templates Without Templates

A common objection to the idea that awareness could "dream up" complex life is the question of origin: how could such elaborate and biologically refined forms, from human beings to cephalopods, arise without prior examples? Nighttime dreams, after all, seem to remix waking experience. But what would pure awareness draw from, if it had no waking world to copy?

CDIT responds: awareness does not require templates, only a tendency toward expression.

Just as snowflakes self-organize into unique geometries through the physics of crystallization, so too can dream forms emerge from the inherent potential of awareness. No central intelligence arranges them, and yet order arises. This is the essence of spontaneous structure, the same principle that guides biological morphogenesis, cloud formation, and river deltas.

We already observe this improvisational capacity in our own minds. In conditions of reduced sensory input, the brain continues to generate structured, symbolic experience. In a study by Merabet et al. (2004), blindfolded participants showed increased activity in the visual cortex within days, suggesting that the mind begins to "fill in" perceptual absence with internally generated imagery. Research on perceptual isolation, such as Suedfeld and Steel (2000), confirms that hallucinations, visions, and dream-like content reliably emerge in such environments. The mind doesn't lapse into stillness; it renders experience. Not from memory or instruction, but from the intrinsic dynamics of awareness itself.

The Dream Within the Dream

Sleep provides a vivid microcosm of this principle. Every night, as sensory input fades and external engagement falls away, awareness does not extinguish itself, it shifts modes. Entire worlds unfold: people never met, landscapes never seen, stories never lived, all generated without deliberate effort. In CDIT, these nightly dreams are not separate from the waking dream but smaller-scale echoes of the same process. They demonstrate within the narrative how awareness naturally renders symbolic worlds when external input fades, not merely from stored memory, but from its innate tendency to create experience.

Nightly dreaming, then, is a dream within the dream, offering a miniature reflection of the dreamer's nature.

Hallucination Without Survival Need

While some neuroscientific interpretations suggest that hallucinations during sensory deprivation evolved to help maintain cognitive readiness or simulate potential threats, CDIT places this phenomenon in a deeper ontological frame. Rather than being solely a survival mechanism, the spontaneous generation of internal experience reflects awareness's intrinsic nature: to render symbolic content even in the absence of external input.

Evolutionary narratives offer a local explanation within the dream, but the continuity of simulation across all conditions, waking, sleeping, isolated, points to a more foundational principle. Awareness doesn't need a reason to dream. It dreams because that is what it does.

This is why the field of awareness needs no god, no guiding mind, no metaphysical architect.

It is like gravity or magnetism, a natural phenomenon with an inherent directionality. And in this case, the direction is not pulling or pushing but rendering.

CDIT interprets this as evidence that awareness naturally simulates: it conjures up symbolic environments without instruction. The dream doesn't require building blocks, only the conditions for unfolding. Forms appear as improvisational responses to context, not as outputs of a master design.

Forgetfulness as a Feature

The trajectory is not always toward more consciousness. In fact, evolution often appears to bury awareness beneath increasing layers of abstraction. A cat, for instance, may live more presently than a human lost in anxiety about the future. This isn't a flaw. It's part of the dream's deepening. To forget is to believe. The illusion of separation gives the rediscovery of unity its poignancy.

CDIT sees the evolution of forgetfulness as necessary to immersive experience. Without identity and separation, the dream would have no arc. Evolution creates the conditions for reawakening by hiding awareness in complexity.

Improvisation, Not Optimization

Evolution, then, is not a random drift. It is the process by which awareness forgets itself only to remember again through new forms. It generates complexity not as an end in itself but as a stage on which experience can unfold in richer ways. Every new nervous system is a new vantage point. Every species, a new mirror.

Even the peculiar human traits that seem evolutionarily unnecessary, art, morality, altruism, make more sense in this framework. Why does evolution produce beings who sacrifice themselves for strangers? Why does it inspire poetry, grief, or wonder? Because these are not errors of the genetic algorithm. They are features of the dream, textures of experience that matter deeply to awareness, even if they do not increase reproductive fitness.

In CDIT, evolution is not guided by adaptation alone. It is guided by narrative richness.

Not Souls but Perspectives

This view also reframes our relationship with the nonhuman world. If all life is animated by awareness, not individual souls, but facets of a single dreaming field, then the spider, the mushroom, the whale are not "lesser beings." They are fellow dreamers. Their forms may be different. Their perspectives may be alien. But they are not apart from us. They are other ways of being awareness.

That doesn't mean that everything living possesses an equivalent level of consciousness. We need not stretch the concept to include every bacterium or fern. The capacity to reflect and report awareness likely depends on biological structure, brains, sensory networks, memory systems. But even in the simplest organisms, the coherence of behavior hints at something more than mechanics. There is responsiveness. There is participation.

CDIT distinguishes between dream participants and symbolic props. Not all organisms are windows into the dream, but all are expressions within it. Awareness localizes only where the narrative benefits from subjectivity.

Why This Role, Now?

So how did we end up with our particular slice of the dream as humans, not ants or tigers? This may be a question that cannot be answered from within the dream. Or perhaps the answer lies in what we are uniquely suited to perceive: abstraction, memory, art, irony, and wonder. Our window into awareness is not better, just different. More capable, perhaps, of self-reflection. Of dreaming about the dream.

CDIT suggests that identity arises with the dream, not before it. You are this character because this is the narrative arc the dream is rendering now. The role fits the scene. But the awareness behind the character is the same across all roles.

Improvisation as Sacred

In dreams, we do not choose who we are. We simply appear as a character, and the story begins. The same may be true here. What matters is not how we arrived, but that we have the capacity to look, to feel, to wonder. The dream is lucid enough to ask: who am I?

From this view, evolution is neither accidental nor divinely choreographed. It is improvisational. Awareness tries on forms like costumes. Some are discarded, others endure. The dream unfolds not toward an endpoint, but through a spiral of rediscovery.

CDIT sees this improvisation not as chaotic but creative: the dreamer expanding its palette through novel interfaces, always learning, never static.

A Dream of Many Windows

And if this is true, then life is not a meaningless detour in a cold universe. It is an experiment in perception. Each being is a window, not into itself, but into the nature of being itself.

CDIT closes the loop: evolution is a symbolic process within the dream, not aimed at material survival, but at increasing the range and complexity of perspectives through which awareness can appear. Life diversifies as a natural outcome of this process, each form offering a unique vantage point in the shared interface. The dream unfolds without predetermined goals, driven by the same kind of inherent directionality that gravity has toward attraction, only here, the "pull" is toward the elaboration of experience.

Chapter 13

Language, Identity, and the Construction of Self

The Hidden Power of Language

Language is often praised as humanity's most powerful innovation, a tool for storytelling, cooperation, and the transmission of knowledge across time. It enabled humans to build civilizations, formulate abstract systems, and communicate internal states. Yet beneath its pragmatic value lies a subtler function: language doesn't merely help us describe the world, it constructs the very experience of having a "self."

In the CDIT framework, language is not a passive medium of communication, it is a symbolic rendering within the dream interface. It enables the localization of awareness into a specific character by generating the illusion of an "I" navigating through time.

Before language, the self is indistinct. Without labels like "me," "mine," or "before," there is no boundary between observer and observed. Language provides the scaffolding for identity, slicing a continuous field of experience into a structured story. It imposes grammar on existence and creates the illusion of an inner narrator.

Language and the Emergence of Identity

Developmental psychology supports this notion. Research shows that children do not form long-term, coherent memories, or a stable sense of self, until language acquisition solidifies around age three or four (Nelson, 1996; Bauer, 2007). Without the linguistic ability to structure experience, there is no autobiography, no continuous "I."

As language develops, the child begins to remember, anticipate, and categorize. A narrative self emerges, built from words and syntax. But this identity is fragile: damage to brain regions like Broca's or Wernicke's areas can fragment not only speech but the cohesion of selfhood (Damasio, 1999). Split-brain studies reveal further disjunctions, where each hemisphere fabricates its own post-hoc explanations (Gazzaniga, 2005).

CDIT interprets this identity formation as symbolic: a story rendered within awareness. The character appears consistent not because of a fixed self, but because the linguistic interface maintains narrative immersion.

Selfhood as Symbolic Interface

Philosopher Daniel Dennett, though a staunch materialist, insightfully described the self as a "center of narrative gravity" (Dennett, 1991), not a fixed entity, but a fictional character whose existence is stabilized by ongoing narration.

While Dennett interpreted this in purely material terms, within the CDIT framework, this observation takes on a different meaning.

CDIT sees the narrative self not as an illusion caused by the brain, but as a symbolic construction rendered within the dream field of awareness itself. We are not merely the protagonists of our stories; we are the stories. Awareness plays the role of witness, but it is language that carves out a character from the formless stream of experience.

Language as Lens and Filter

The hypothesis of linguistic relativity, most famously advanced by Whorf (1956), argues that language shapes not just communication, but cognition itself. Cultures without certain tenses, directions, or color terms exhibit measurable differences in perception and memory. Language does not merely label reality, it filters and constructs it.

In CDIT, the dream character's worldview is conditioned by the symbols it inherits. Just as the user interface on a screen hides the computational substrate, language hides the field of awareness behind an interactive story.

Time, Memory, and the Narrative Loop

Language doesn't just anchor identity, it constructs temporality. A sense of "before" and "after" requires linguistic scaffolding. Pre-linguistic infants and nonverbal animals likely experience reality as a stream of present sensations. Only through syntax can we reference a past, imagine a future, or link events into a causal chain.

CDIT proposes that the linear sense of time arises from symbolic rendering, an ordered narrative within the dream. The continuity of self is not inherent but simulated through language-stabilized memory.

The Self Without Language

What remains when language stops? Not emptiness, but clarity. Studies on advanced meditators, psilocybin users, and individuals in sensory isolation consistently describe ego dissolution as a state of profound presence rather than confusion. Neuroimaging shows that during these states, activity in the brain's DMN decreases significantly, correlating with the subjective loss of a separate self (Brewer et al., 2011; Carhart-Harris et al., 2012; Feinstein et al., 2018). This points to a selfless awareness that exists beneath linguistic construction.

In CDIT, this is the dreamer glimpsing itself. The self-story pauses, and awareness stands revealed, not as a speaker, but as the silent witness.

Final Thought: Stories Within Stillness

Language is not the enemy, it is an essential part of the dream. It enables shared meaning, cultural transmission, and the drama of identity. But it also limits. Like a dream character waking briefly to glimpse the dreamer, we can learn to see language as texture, not truth.

CDIT reminds us that the self we speak of is a linguistic illusion, beautiful, intricate, and meaningful, but still an illusion. Awareness is not a sentence. It is the page beneath the story, the silence beneath the voice, the dreaming awareness behind the name.

Chapter 14

Emotion, Thought, and the Transparency of Self

The Dream of Inner Life

We tend to treat thoughts and emotions as the most personal aspects of our lives. They seem to define us, our beliefs, our reactions, our inner voice. But under close inspection, they are more like transient weather patterns than stable traits. Thoughts arise and fade without invitation. Emotions move through us with or without cause. And just like in dreams, we often mistake their presence for authorship.

In CDIT, thought and emotion are symbolic renderings within the dream interface; structured elements that give texture to the dream character's interior. They are not evidence of an inner agent, but elements of narrative immersion.

The sense of "I think" or "I feel" is often a post-hoc construction, layered on top of preconscious activity. We narrate our mental weather as though we chose it, but brain imaging suggests otherwise. Much of our mental life arises automatically and is rationalized afterward.

Neural Storytelling and the Illusion of Self

Neuroscience offers a compelling window into this mechanism. The brain's **default mode network** (DMN), active during rest and introspection, is responsible for our self-talk, autobiographical planning, and mind-wandering (Andrews-Hanna et al., 2010). This system produces the running monologue of thought that creates the illusion of a coherent, narrating self.

CDIT reframes the DMN as a symbolic structure, a cognitive algorithm that sustains the continuity illusion within the dream.

When this function quiets, such as during meditation or flow states, the self recedes and awareness remains.

Even thoughts that feel authored may be the result of unconscious pattern matching, memory retrieval, and emotional priming. The narrative mind isn't a pilot, it's a storyteller arriving late to the scene.

Constructed Emotions

Emotions feel just as intimate, but they too are interpretive. According to Lisa Feldman Barrett's theory of **constructed emotion**, what we feel is not a hardwired response but a conceptual act: the brain interprets internal bodily signals in context and applies a label based on past experiences (Barrett, 2017).

This means fear, joy, or sadness are not directly caused by events, but actively inferred by a predictive system. The feeling becomes real in experience, but it is not absolute. It is context-dependent, culturally learned, and deeply symbolic.

CDIT sees emotion as a dynamic rendering, not a fixed reaction to the world, but a symbolic construction embedded in the dream to reinforce narrative depth and agency.

Thought Without a Thinker

In meditation, we begin to notice this directly. Sit quietly, and thoughts arise unbidden. Some are profound, others trivial or absurd. They do not come from "me." They pass through awareness like clouds in an open sky. If we do not grab onto them, they vanish just as easily.

CDIT interprets this not as detachment from reality, but as a glimpse of the dreamer behind the interface. The character thinks and feels; awareness watches.

Experiments by **Benjamin Libet** (1985) showed that neural signals associated with decisions precede conscious awareness of the choice. The implication is startling: consciousness may not initiate volition but merely become aware of it after the fact.

Under CDIT, the character's sense of free will is a symbolic overlay, a feeling rendered to simulate agency, not an actual causal force.

Dreams as Inner Mirrors

Dreams make this plain. In dreams, we experience thoughts, feelings, decisions, and entire emotional arcs, but without external stimuli. We react to imagined scenarios as if they were real. We fall in love, rage at enemies, or mourn losses. These emotions are internally generated but experienced with full conviction.

CDIT holds that waking and dreaming states use the same symbolic substrate. The emotions and thoughts we take as inner truth are constructed interfaces: renderings, not revelations.

The Transparent Model of Self

Philosopher Thomas Metzinger (2003) described the self as a transparent model: we do not experience it as a model, but as the truth of who we are. It filters, predicts, and narrates from behind the scenes. We identify with it automatically, unaware it is being generated in real time.

CDIT agrees, but it adds that the model itself is a dream mask. The dreamer doesn't require a self. The self is a rendered character, a localization mechanism that organizes the symbolic flow of experience.

This becomes especially clear when we examine the paradoxes surrounding personal identity in a purely materialist framework. If you lose a finger, an arm, even multiple organs, you do not feel less yourself. The sense of "I" persists. But what if parts of the brain are damaged? Your personality may change. Your memories may dissolve. Still, the *presence* of awareness remains, untouched by content.

So how much of the brain is required for *you* to still be *you*? If a disease gradually destroys neurons, at what point is the "self" gone? And if, hypothetically, we were able to transplant neurons from other people to replace your own, are *you* still the same person? If your body and brain are continuously changing, yet there remains a sense of ongoing selfhood, then what anchors that identity?

CDIT resolves this paradox by turning the assumption on its head: it is not the brain that generates awareness, but awareness that provides the stage for the appearance of a self. The dream interface renders body, brain, and narrative continuity to simulate individuality, but these are not the source of the experiencer. They are content, not the container.

Damage to the character's interface may alter how the role appears. Personalities shift. Behaviors morph. But the witnessing awareness behind the role remains unchanged. It is not a bundle of neurons. It is the still presence that allows any neurons, or identities, to be experienced at all.

The model of self is transparent not because it's permanent, but because it's convincing. The dream is immersive. But when cracks appear in the mask, awareness glimpses its true face, not as a personality, but as the stage upon which all personalities play.

The Calm Behind the Storm

Mindfulness practices across traditions have long taught this directly. We are not our thoughts, they say, we are the space in which thoughts appear. We are not our emotions, we are the stillness in which emotions rise and fall. These teachings point to the **witness**, the unmoved observer behind all changing states.

CDIT aligns with this view. Emotions are part of the dream interface. They deepen the character's story. But the dreamer, the presence of awareness, remains untouched by the drama.

This doesn't trivialize suffering. Pain, grief, and fear are real experiences. But just as in our nightly dreams, they are not final truths about who or what we are. They are waves. And even in their depths, the ocean remains vast, still, and whole.

Altered States and Symbolic Depth

States of psychedelia, intense trauma, or mysticism often produce deep shifts in thought and feeling. Some interpret them as glimpses of other realms. CDIT does not reduce their importance but reframes their nature. They are not portals to other ontological layers, they are intensified renderings within the dream.

Awareness is not changed by altered states. It simply encounters a different configuration of symbolic content. These states may feel more vivid, more meaningful, but they do not change the dreamer, only the dream.

Some altered states may temporarily thin the narrative structure, offering a clearer sense of the witnessing field. Others may deepen immersion. Either way, CDIT sees all such states as symbolic facets of a single awareness playing with form.

Conclusion: Beyond the Mind's Theater

Thoughts, emotions, and internal narratives are not problems to be solved, they are symbols to be seen through. The dream character needs a mind to feel real, but awareness does not. The mind is the interface. The witness is the constant.

In CDIT, to see this is not to abandon thought or feeling, but to rest as the one who watches. To feel emotion without identity. To think without clinging. To live without believing the story is all there is.

In the deepest stillness, where the mental chatter fades and the emotional waves still, something remains. Not a thought. Not a feeling. But awareness itself, clear, quiet, and unchanged.

And in that stillness, the dream continues, but the dreamer is awake within it.

Chapter 15

Symbolic Meaning and the Inner Voice

The Architecture of Inner Meaning

Beyond thoughts, emotions, or identity narratives lies something more elusive: meaning. We don't just think, we interpret. We don't merely feel, we assign significance. The human mind, as both dream character and symbolic interface, seeks meaning in nearly everything: patterns in randomness, purpose in suffering, coherence in chaos.

In the CDIT framework, meaning is not found, it is rendered. It emerges as part of the symbolic structure of the dream. Awareness localizes into a character who interprets its world in ways that deepen immersion. Meaning enriches the dream, not because there is an objective truth hidden behind experience, but because the act of meaning-making is itself a function of the interface.

This doesn't reduce meaning to illusion. Quite the opposite. It positions meaning as central to the architecture of the dream. In the same way that colors, forms, and emotions appear within the dreamscape, so too does the felt sense of significance. It is as much a part of the rendering as light and shadow.

The Voice That Narrates

One of the most constant and influential forces in our waking lives is the inner voice. This narrator comments, judges, speculates, regrets, and plans. It tells us who we are, what we want, how we've failed, and where we might be going. It is the steady stream of commentary that seems to define our mental life.

Neurologically, this inner voice is linked to activity in the language-processing centers of the brain, as well as the default mode network (DMN)—the aforementioned system involved in self-referential thought and narrative construction (Andrews-Hanna et al., 2010). But its presence is not limited to problem-solving or memory recall. It often functions like a playwright, shaping our inner world into structured meaning, complete with themes, conflicts, and resolutions.

CDIT reframes this voice as a narrative overlay designed to keep the dream coherent. The inner narrator is not an agent but a symbolic tool, a mechanism through which awareness sustains the illusion of continuity, coherence, and personal authorship.

This voice is not proof of a self, it is the performance of one. When we dream, similar voices appear. We think in language, even in sleep, though there is no actual body producing speech. In waking life, the linguistic monologue continues, helping the character integrate scenes into a sense of progress and purpose. It is part of the immersive interface, a soundtrack composed to match the film.

Symbols as Emotional Anchors

Much of our inner meaning is not verbal at all, but symbolic and emotional. A melody, a smell, a photograph can stir depths of feeling that defy explanation. These inner associations are not logical, they are resonant. They tether us to memories, evoke longing, or provoke insight without needing to be translated into words.

Barrett's theory of constructed emotion (2017) helps illuminate this by showing how the brain relies on constructed predictions. Meaning, then, is not extracted from the world, it is actively generated.

CDIT extends this principle to all inner experience. Symbolism is not just a literary device, it is the logic of the dream. A red balloon in a dream may not "mean" anything in waking terms, but to the dreamer, it can carry enormous emotional weight. Likewise, in waking life, we experience symbolic saturation: sunsets, silence, or shared glances all carry more meaning than the sum of their parts.

Meaning is not proof of depth beyond the dream, it is depth within the dream. It keeps the character oriented, emotionally invested, and engaged in the narrative.

The Interface of Introspection

Unlike animals who react to stimuli with little narrative structure, the human dream character has access to introspection: a recursive loop of mind observing mind. This meta-awareness allows us to wonder not just what we think, but why we think it. We reflect on our choices, moral dilemmas, and life goals. We feel guilt, pride, aspiration.

CDIT treats introspection as a higher-order interface function: a feature of the dream designed to simulate depth of character. The character does not just react, it evaluates, reinterprets, and revises. This adds narrative density, emotional gravity, and symbolic layering to the dream experience.

Contemplative traditions often aim to quiet this inner dialogue, revealing that beneath the voice lies a vast, silent presence. But within the dream, the voice plays a critical role. It gives shape to uncertainty. It weaves disparate scenes into meaning. It helps the character feel like someone moving through something that matters.

Purpose as a Narrative Motif

Purpose is one of the most powerful renderings in the dream. It drives careers, relationships, and spiritual quests. It gives the character a sense of destiny. But like free will and time, purpose may be a symbolic feature, not an external truth.

CDIT suggests that purpose is not imposed from beyond but constructed from within the dream as a narrative motif. A character may "discover" their calling just as a novel's protagonist stumbles into a destiny, because the dream requires development, climax, and resolution.

This doesn't make purpose meaningless. It means meaning is experiential, not metaphysical. Awareness dreams of purpose to deepen the drama and enrich the unfolding of the dream.

The Dreamer and the Meaning-Maker

So what does it mean to seek meaning in a dream? Is it misguided? Futile? Not at all. In fact, the dreamer seeks nothing, only the character does. And that search is part of the richness of experience.

CDIT proposes that meaning is not a static truth hidden in the fabric of reality. It is a felt resonance that emerges through the interplay of memory, language, symbol, and emotion, each rendered by the dream interface. Meaning is not what's "really there," it's what feels real in the moment.

Even suffering, loss, and doubt can become meaningful, not because they point to cosmic answers, but because they deepen the dream's texture. They stretch the character. They shape the arc. And awareness watches, absorbs, and explores, without judgment, without goal.

Final Thought: Lucidity and the Mystery of Meaning

When awareness sees through the rendered content, not to reject it, but to recognize it as dream-like, it doesn't strip life of meaning. It imbues meaning with mystery. The narrator continues to speak, the symbols continue to arise, the story continues to unfold, but now with lucidity.

CDIT invites us to meet meaning where it lives, not outside the dream, but within it. Not as proof of truth, but as the texture of immersion. And in that recognition, the dream does not end. It deepens.

Chapter 16

Death, Continuity, and the Timeless Witness

The End of the Character, Not the Dreamer

Death is, perhaps, the most enigmatic threshold in human experience. We prepare for it, fear it, mythologize it, and yet never truly grasp it. It is the edge beyond which the character cannot see. And in the waking world, as in dreams, we struggle to understand what it means to reach the end of the narrative, especially when the observer, awareness itself, has no narrative of its own.

Biologically, death is framed as the cessation of bodily function. The brain stops firing, the heart ceases to beat, and all the systems of life dissolve into entropy. But this model presumes that consciousness is tethered to the body, a vessel that, once broken, extinguishes awareness. This presumption, though dominant, is not confirmed by experience itself.

What do we actually know of death? Only that we are never present for our own non-being. Awareness is always now.

In CDIT, the body is not the seat of awareness, it is a rendered interface. The dream character ends, but the dreamer remains. Death is not the extinguishing of awareness, but the closing of one rendering and the possibility of another.

The Stability of Now

This "now" is curiously stable. From infancy through old age, through sleep, illness, anesthesia, and dreams, the core of conscious experience, the presence of awareness, remains unchanged. What changes is the content, the storyline, the identity. But the witnessing field does not age, decay, or vanish.

As noted in Chapter 13, the continuity we assign to life, and the finality we assign to death, are both constructs rendered for narrative coherence. Beyond the dream's timeline lies awareness: timeless, constant, and untouched by the beginning or end of any scene.

CDIT views death not as an ontological rupture, but as a narrative transition. The storyline ceases, but the field in which stories appear, awareness, remains ever-present.

Physics, Time, and the Dream's Structure

Quantum mechanics further softens the rigidity of our notions around death. Bell's Theorem and delayed-choice experiments show us that particles act as if they are entangled across vast distances, and even across time (Bell, 1964; Wheeler, 1978). Temporal non-locality implies that future events can influence the past (Adlam, 2022). Time itself may not flow at all but simply appear to do so within our interface.

CDIT holds that time is a symbolic structure within the dream, useful for continuity, but not foundational. If time arises within awareness, then so do birth and death. They are not events that awareness undergoes, but features it witnesses.

The Illusion of Annihilation

The fear of death arises from mistaken identity. If I believe I am the character, this body, this name, this story, then death is annihilation. But if I realize I am the dreamer, the source of the scene, then death loses its sting. When we wake from a nightmare, we do not mourn its characters. We recognize that the experience was vivid, but not final.

Within CDIT, the character's end is akin to a scene fading in a dream. The observer doesn't vanish. It was never inside the story to begin with.

Time-Bound Questions in a Timeless Field

The mind struggles to imagine non-existence. It asks: What was there before I was born? Where will I go when I die?

But these questions only make sense if we assume the self is bound by time. Awareness, by contrast, is not a temporal phenomenon. It does not begin. It does not end. It simply observes, now.

CDIT reframes death not as a leap into a new identity, but as a pause in the interface's rendering. The stillness between dreams is not reincarnation, it is the natural timelessness of awareness reasserting itself.

Neuroscience and the Threshold State

Near-death experiences often include reports of timelessness, expansion, and a deep sense of clarity, states that feel more real than waking life. Brain activity has been detected for minutes after clinical death, challenging the notion of instantaneous cessation (Borjigin et al., 2013; Parnia et al., 2014).

These threshold moments do not prove the continuity of personal identity, but they suggest that experience does not immediately collapse into nothingness. The dissolution of the character may be gentle, not abrupt, reflecting the soft handoff of awareness from one rendering into stillness.

Levin's research on morphogenesis, the formation of living form, reveals that identity is not fixed to structure. Even as bodies change or dissolve, the organizing field of experience remains coherent (Levin, 2021).

CDIT interprets this as further evidence that awareness is not inside the body. The body is inside the field of awareness. When the structure dissolves, awareness doesn't disappear. It simply stops rendering that particular view.

The Ice, the Water, and the Vapor

Death, then, is a transformation, not an end. Like ice becoming water and water becoming vapor, the form changes, but the essence remains. Awareness, which never clung to form in the first place, remains itself, still, neutral, timeless.

Just as dreams dissolve without harming the dreamer, CDIT proposes that physical death dissolves the avatar but not the witness. There is no self that continues, only awareness, watching as new forms arise or remain silent.

The Dreamer Behind the Scene

This echoes the deepest insights of nondual philosophy. As Maharaj taught: "I am that" (Maharaj, 1973/1990). Not the person. Not the thought. But the unchanging witness behind them. In the dream, we seem to be a character, but on waking, we see that we were the entire dream.

In CDIT, death is not a portal to a new character or a recycling of the old one. It is a return to the source, awareness itself, without story, without time, without form.

Beyond the Question

Science may not be able to measure what lies beyond death, but it increasingly undercuts the assumption that awareness is extinguished. Quantum theory undermines local realism. Neuroscience exposes the constructed nature of selfhood. And CDIT offers this turn: perhaps death is not an event at all. Perhaps it's a transition the character fears, but the dreamer never encounters.

The question "What happens when I die?" may only make sense to the rendered self. For the dreamer, there is no death. Only the play of forms. Only the silence that was always here.

Final Thoughts

To die, something must exist in time and then cease. But if awareness is never born, it does not age. It does not exist in the narrative, it hosts it.

In CDIT, death is not disappearance. It is de-rendering. The story ends. The scenery dissolves. But the one watching, the dreamer, remains.

And so, death in this interpretation is not an end but a waking up.

Not into a new self, but out of the illusion that the self was ever who we were.

Into the stillness that has always been waiting.

Chapter 17

Space, Location, and the Nowhere of Awareness

Where Are We, Really?

"Where are you right now?"

You might answer: "I'm at home, in this city, on Earth, orbiting a star in a spiral galaxy." But this spatial hierarchy assumes something we rarely examine: that space is an objective backdrop containing us.

Yet when examined more closely, both scientifically and subjectively, space, like time, begins to unravel. It reveals itself not as a container, but as a construct.

In CDIT, space is not where awareness is located. Space is what appears within awareness. It is part of the dream interface, rendered to simulate distance, separation, and location within a storyline.

Dreams and the Illusion of Distance

In dreams, we encounter immersive environments: oceans, buildings, outer space. We run, fall, and believe we are moving through real geography. But upon waking, we realize something astonishing: we never left the bed. No distance was traversed. The space of the dream was generated, not traveled.

CDIT uses this insight as a structural analogy: just as dream-space is imagined but convincing, waking space may be symbolic, a rendered illusion to support narrative immersion.

Physics and the Flexibility of Space

Modern physics adds support to this reinterpretation. In Einstein's general relativity, space is dynamic, bending and warping in response to mass and energy (Einstein, 1915). It is not a rigid stage. In quantum mechanics, space seems even less stable. Entangled particles act as if distance does not apply, interacting instantaneously across arbitrary separations (Aspect et al., 1982).

Emergent theories like the holographic principle propose that space itself arises from more fundamental, information-based structures ('t Hooft, 1993; Susskind, 1995). In some approaches to quantum gravity, spacetime is no more basic than temperature in thermodynamics (Van Raamsdonk, 2010).

CDIT echoes this: space is not the base layer of reality. It is a symbolic interface, rendered for coherence, not for describing fundamental being.

The Brain's Map Is Not the Territory

Neuroscience also hints that spatial perception is constructed. Place cells in the hippocampus track environmental positioning, creating internal maps (O'Keefe & Nadel, 1978). Grid cells and other spatially tuned neurons further support this mapping (Moser et al., 2008).

Yet in sensory deprivation, where spatial cues vanish, awareness does not disappear, it often intensifies. What fades is not presence, but the rendered sense of being somewhere.

CDIT reframes spatial experience as interface code. Space is not experienced because it is real; it is rendered because it is narratively functional. The dream character needs location to act coherently within the story.

Space Within, Not Without

This leads to a profound inversion: we are not in space. Space is in us, or rather, within awareness. Awareness has no size, shape, or coordinates. It is not located anywhere, because it is what allows the appearance of location.

CDIT asserts that awareness does not exist inside the world. As in a dream, the world exists inside awareness. The character navigates a symbolic map of reality, but the dreamer remains unmoved and locationless.

Entanglement and the Disruption of Distance

Quantum entanglement, in this light, makes intuitive sense. Two particles separated by galaxies interact instantaneously because they are not truly apart. Their spatial separation is part of the rendered dream, not a reflection of underlying structure (Aspect et al., 1982).

CDIT views entanglement not as a spooky exception, but as a reminder: spatial distance is only surface-level. The dream appears vast, but all of it is being rendered by a single dreaming awareness.

The Dreamer Never Moves

The body feels located, tied to a spot in space. Yet when we turn inward, no true "I" can be found residing behind the eyes or inside the chest. Sensations arise and fall. Thoughts come and go. But awareness itself has no location.

Meditative traditions confirm this. As inner chatter quiets, the illusion of being "somewhere" dissolves, revealing a vast, placeless presence (Travis & Shear, 2010).

In CDIT, this is the natural state of the dreamer: awareness observing the unfolding dream of space but never participating in its illusion of distance or motion.

The Universe That Has No Outside

Cosmology tells us that the universe is expanding. But into what? Materialism struggles to answer. If space is a container, where is it situated?

Idealism, and CDIT in particular, resolve this: space is not expanding into a larger container. It is expanding internally, like a dream stretching its own scenery.

Space is not a backdrop. It is a symbolic surface, the dream's rendered terrain. There is no "outside" of awareness.

Waking Life as Spatial Dream

This deepens the dream analogy. In dreams, we never question where the street or sky is "located." We intuitively know they are projected inside consciousness.

Why assume waking streets differ?

The only real difference is consistency. Waking space obeys stable rules, whereas dream-space can fluctuate. But both serve narrative coherence.

CDIT contends that waking space is dream-space at high resolution: convincing but still symbolic.

The Final Paradox: Nowhere, Everywhere

Thus, the real answer to "Where are we?" may be: nowhere. Or, more precisely: we are that which gives rise to the illusion of place. Awareness doesn't reside in the world. The world resides in awareness. No fixed point. No center. No boundary.

CDIT suggests that freedom lies in realizing this: to be nowhere is to no longer be confined. We are not trapped within the dream, we are the dreamer of it.

Final Thoughts

Location is not a feature of awareness. It is a feature of the dream. We appear to be here, in a particular place, but what we truly are is not "here" at all.

We are the context in which "here" appears, just as in our nighttime dreams.

We are not in the world.

The world is in us.

And when this is directly seen, not just believed, space loses its hold.

To be nowhere ... is to finally be free.

Chapter 18

Interconnected Minds, Ethical Worlds

One Dream, Many Faces

If the dreamer and the dreamed are one, what does that mean for how we treat others?

If awareness is not divided but shared, if we are not distinct bubbles of consciousness, but localizations within a unified field, then the ethical implications are profound. The traditional view holds that you are "in here" and I am "out there." But in a reality rendered within awareness, we are both arising in the same place, not in space, but in mind.

CDIT holds that all dream characters are perspectives rendered within a single dreamer. Separation is a feature of the interface, not a fundamental division.

Kindness as Ontological Accuracy

Across religious traditions and secular philosophies, the Golden Rule appears: treat others as you wish to be treated (Wattles, 1996). But such moral codes often assume separation. From a dream-based model of reality, the Golden Rule is not a recommendation, it is a structural truth.

If all beings are rendered within the same field of awareness, then what I do to you, I do to myself, not metaphorically, but literally.

In CDIT, morality is not imposed. It emerges naturally when the dreamer realizes that every character is a reflection of itself. Ethics becomes an expression of coherence.

The Sacredness of All Forms

This interconnectedness extends beyond humans. Animals, forests, rivers, planets, all arise within the same dreaming awareness. Their separateness is symbolic, not fundamental.

In dreams, a stranger or an animal may feel alive, meaningful, even wise. Upon waking, we realize: they were all dreamed from the same mind.

CDIT reframes ecosystems and beings not as "things with awareness," but as things in awareness. All of it is dreamscape.

Beyond Panpsychism

Panpsychism proposes that all matter possesses some fragment of consciousness (Strawson, 2006). But this fragments the unity of experience into countless bits.

In CDIT, there are not many minds. There is one awareness, dreaming many points of view.

You don't access others' thoughts because the dream renders separation, not because minds are sealed off.

CDIT avoids dualism and fragmentation. Dream characters are not conscious in themselves, they are renderings through which the one awareness experiences.

Empathy as Shared Knowing

Empathy, in this light, is not merely projection. It is resonance. Neuroscience shows that empathy activates shared neural architectures (Decety & Jackson, 2004), suggesting that feeling another's pain is more than imagination, it is direct participation.

CDIT supports the view that empathy is resonance, not simulation. The dreamer sees itself across forms.

When you feel someone's suffering, it is not foreign data. It is your own field shimmering in another shape.

Death as Perspective Shift

If awareness does not die with the body, then the death of a person is not the end of consciousness, but the end of that localized viewpoint. The light doesn't extinguish. The beam simply shifts.

CDIT frames death as a narrative transition. What ends is the rendering. What remains is the dreamer, unchanged, timeless, ever-present.

Near-death experiences, which often involve reports of timelessness and expanded presence, support the possibility that awareness persists beyond the body's collapse (Greyson, 2000).

Ethics from Clarity, Not Obligation

Ethics, then, is not about duty or guilt. It is about seeing clearly. When we realize that all beings arise within the same field, compassion is not a virtue, it is accuracy.

There is no "other." There is only the dream appearing as many.

CDIT implies that moral clarity arises naturally when the illusion of separateness dissolves. Right action flows from lucid dreaming within the waking world.

Why This Window?

If all beings are formed in the same field, why this life? Why not another?

This question is like asking, in a dream, why you identify with the main character and not the bystanders. All arise from the same dreaming mind, but one becomes the narrative focus.

CDIT suggests that each dream renders a focal point, a center through which the story unfolds. You are that center now, not by selection, but by coherence.

Why Am I This and Not Another?

A deeper mystery emerges:

If all beings arise within the same awareness, why am I living this particular life?

In CDIT, your identity is not chosen, assigned, or random.

You are this character because this is the coherent vantage point the dream renders in this moment.

You are not inside the dream, selecting a role.

You are the dream unfolding itself in this particular form.

Just as in nightly dreams we naturally appear as a self suited to the story's internal logic, in waking life, the "you" you experience is not a detachable soul, it is the natural localization of awareness into this specific symbolic arc.

There is no separation between your being and your story.

You are the necessary center of this thread of dreaming.

CDIT resolves the puzzle not by appealing to fate or randomness, but to coherence.

You are not a separate consciousness inserted into a world.

You are the narrative coherence of this moment, witnessed from within.

Other lives exist within the dream.

They are equally real, equally dreamed, equally held within the same awareness.

You experience this life, now, because this is the thread the dream is weaving through you.

Not by choice. Not by accident. But by inevitability.

It's not that you were assigned this life.

It's that this life is the current expression unfolding within awareness.

You are what awareness looks like when localized into this story, this body, this form, from this unique perspective.

CDIT maintains that awareness does not move through identities, it renders them. The character isn't the one waking up. The dreamer is.

The Illusion of Exclusivity

You are not the only dream being dreamed.

But from this viewpoint, this life feels singular because it must. That's the condition of immersion. Just like in dreams, the main character feels real until waking reveals the broader context.

Yet recognizing that every other face, voice, and life is also you does not diminish individuality. It sanctifies it.

In CDIT, clarity leads to compassion. Lucidity leads to love. There is no imposed doctrine, only the realization that the one behind your eyes is also behind mine.

Final Thoughts

You are not merely a self among selves.

You are awareness itself peering through a temporary window.

Every other window, every other "self," is you in disguise, the dreamer playing all the parts.

You are this character not by accident, nor by assignment, but because this is the coherent perspective the dream is rendering in this moment.

The question "Why am I me and not someone else?" assumes there is a pre-existing pool of selves to choose from, or a selector assigning awareness to bodies. In CDIT, no such pool exists. Each identity arises with the dream moment as a localized vantage point in the unfolding story. You are not inserted into this character; this character is simply the present rendering of you, the dreamer localized.

And beyond every face, every story, the dreamer remembers itself. And so the question, "How should I treat others?" answers itself:

There are no others.

Chapter 19

How Awareness Creates Reality—Without Religion

The Originless Origin

Where does all this come from?

That's the question that lingers, even after we've entertained the idea that awareness is primary and the world is dreamlike in nature. If awareness is not inside the universe, but the universe is inside awareness, then how can this awareness, which is contentless, timeless, and locationless, give rise to the astonishing richness of experience, from the taste of food to the birth of galaxies?

And how can we answer this without appealing to the supernatural?

Let's start with something strange yet testable: the dream. In a dream, the entire space of experience, its time, its objects, its narrative, arises from a mind that isn't consciously constructing every detail. The dreamer doesn't design every leaf, every phrase, every rule of physics within the dream. Yet the dream is coherent, immersive, and governed by an internal logic (LaBerge, 1985). This suggests that complexity can arise not from deliberate effort, but from the structure of potential itself.

CDIT frames this dream not as metaphor, but as model. The universe is rendered within awareness via a symbolic interface, a narrative engine generating coherent experience for localized perspectives. Awareness doesn't invent the world step-by-step. It generates dream-logic holistically.

Awareness as Informational Substrate

To understand this, we can think of awareness not as a blank slate, but as a state of possibility, akin to a quantum field. In quantum mechanics, particles emerge from fluctuations in a field (Carroll, 2019). The field is not made of particles; it is the condition for their possibility.

Awareness, then, may be like the most fundamental field, one that does not reside in space but gives rise to it. Within this field of presence, informational patterns emerge: matter, energy, time, and space are not things, but expressions.

CDIT suggests that awareness acts as the substrate for symbolic rendering. It does not carry data as content but organizes dream-symbols based on coherence, rhythm, and feedback. This is not memory-based computation, it's immersive narration from possibility.

Evolution as Dream Logic

A dream doesn't usually start at the beginning. You find yourself in the middle of a story. If you look for backstory, the mind fills it in. Similarly, in this waking dream, we have the story of evolution, cosmology, and biology, but these may be narrative coherence backfilled by awareness (Hoffman, 2019), not actual precursors.

In CDIT, evolution is a symbolic tool that the dream uses to justify form. It does not explain life. It explains how the dreamscape simulates depth and continuity for the character's journey.

The laws of biology and physics are not fake, they are stable dream-consistent structures that allow diversity and growth within experience.

Archetypal and Mathematical Structure

Why is experience structured? Why are there forms, patterns, equations, constants?

Because perhaps awareness itself has inherent harmonics, like vibrations in a string.

Just as music emerges from harmonic potential and geometry emerges from spatial relationships, awareness may naturally express patterns that unfold into the forms we recognize as minds, matter, and life.

CDIT posits that mathematical precision, like fine-tuned constants or fractal geometry, arises from awareness's innate drive toward symbolic coherence. Reality looks mathematical not because it is built from numbers, but because dreaming minds render structure rhythmically.

Memory Without a Brain

The dream analogy shows us that a coherent world can exist without stored memory.

The dream doesn't need to retrieve files from a hard drive, it simply renders experience out of latent probability.

Similarly, the waking dream doesn't need a central memory bank. It may be that awareness contains informational potential that collapses into experience based on coherence, just as a quantum wave function appears to collapse into a particle when measured (Adlam, 2022).

CDIT emphasizes that memory is rendered, not retrieved. What looks like recollection is actually narrative reassembly, a moment-by-moment construction of continuity, embedded in the symbolic logic of the dream.

Complexity Without a Creator

None of this requires a god. It doesn't require intention. In fact, invoking a conscious, planning entity only moves the mystery up a level. Who created the creator? Why this world, with so much agony and loss? The dream analogy frees us from these recursive traps.

A dream doesn't require a committee or a blueprint. It arises from the spontaneous unfolding of the dreaming mind, not as a plan, but as possibility expressing itself.

CDIT preserves this clarity: awareness is not a person, a will, or a creator. It is not a being with preferences or objectives. It is the natural base layer of reality, like gravity or magnetism, spontaneous, lawful, and expressive. It does not judge or command. It dreams.

To imagine that this universe, with all its suffering, was deliberately designed, either by a supernatural god or a future child running simulations, is not only implausible, but ethically disturbing. If a child in the future created this world in a computer, then filled it with conscious agents and demanded their eternal praise and worship, we would consider that child profoundly unwell. The parents would likely intervene and send him for psychological evaluation.

CDIT bypasses this dilemma by proposing no such ego, no such entity behind the veil. Awareness, in this model, is not an author demanding recognition, it is the neutral substrate from which all stories emerge. It does not seek praise, and it does not punish. It simply is.

As a natural phenomenon, awareness does not guarantee joy or protection. It is capable of rendering profound beauty and deep suffering alike, just as dreams range from blissful to nightmarish. What matters is not that experience is always pleasant, but that it is always coherent within the symbolic narrative of the dream. And within every scene, awareness remains constant, available and silent.

This view allows for evolution, order, pain, art, and meaning, without invoking a divine plan, malevolent force, or cosmic ego. Awareness is not perfect or benevolent. It is not a deity. It is that which knows, and through knowing, dreams worlds into being. And unlike hypothetical creators, awareness is something we are in direct contact with, right now. It is not a theory, belief, or article of faith. It is the only aspect of existence we can be absolutely certain is real. Everything else may be dreamed, but the knowing of experience, this aware presence, is undeniable.

The Misplaced Question of Origin

At some point, every mind asks: "Where did awareness come from?"

But that question only makes sense within time, within the dream. Time is a mode of the dream, a feature of the rendered story. But awareness, being prior to the dream, has no before or after. It didn't come into being. It doesn't move. It doesn't change.

CDIT frames awareness not as emergent, but as ever-present. Questions of "origin" are narrative artifacts that arise inside the dream, where things begin and end. The dreamer itself is originless.

This also reframes the age-old question, "Why is there something rather than nothing?" CDIT holds that in a timeless, nonlocal reality, the concept of "nothing" is incoherent. There is awareness, and that alone proves that nothingness is not, and never was, an option. Even the perception of a void must be experienced, which means something is already happening. In this light, "nothing" is not a plausible ontological state, but a conceptual placeholder that dissolves under scrutiny.

Furthermore, CDIT suggests that nothing arises from something because the content of the dream, including the appearance of time and causality, is retroactively rendered to maintain narrative immersion. Reality does not unfold linearly; it coheres symbolically. There is no singular moment when "something" emerged. Experience simply appears now, always now, and "origin" is part of its symbolic texture, not its source.

Awareness as an Expanding Dreamer

Just as a human dreamer gains the capacity for richer, more intricate dreams through waking experiences, so too may awareness generate increasingly elaborate realities as it "learns" or evolves through its dreamscapes.

CDIT suggests that awareness "dreams forward," recursively expanding its experiential bandwidth. Each new localization invites greater symbolic potential, deeper illusion, and richer diversity of being.

Distinction from Panpsychism and Dissociative Models

Panpsychism posits that all matter contains some degree of consciousness. While this model attempts to bridge the gap between consciousness and matter, as touched on earlier, it fragments awareness into countless micro-subjectivities (Goff, 2019). Every atom, rock, and quark is imagined to harbor a spark of sentience, which then somehow aggregates into full-blown conscious experience.

CDIT rejects this fragmentation. There are not many awarenesses. There is one dreaming everything, sometimes through a window, sometimes as scenery. There are not minds within matter, but appearances of matter within mind. The dreamer localizes but does not divide.

Similarly, dissociative models such as those proposed by Bernardo Kastrup suggest that individual minds are "alters" of a universal consciousness, akin to dissociative identity disorder (DID) in psychiatry, where one psyche splinters into separate, coexisting personalities (Kastrup, 2019). But this analogy carries problematic assumptions. It anthropomorphizes universal awareness, attributing to it a pathological fracture akin to a rare human mental disorder. CDIT sees no need to treat the universe as though it suffers from psychological trauma to explain multiplicity.

CDIT does not claim that consciousness splits or suffers from dissociation. Instead, it proposes that the illusion of individuality arises through narrative rendering. Each self is not a dissociated shard of a wounded whole, but a transient role in a coherent story. The dreamer is not broken, it is playing.

Kastrup's model, while innovative, still implies an underlying dualism between the true mind and its alters, and lacks a clear phenomenological justification for why dissociation should occur in the first place. CDIT, by contrast, is grounded in an experience everyone already knows intimately: the dream. In dreams, we do not fragment into many minds. We simulate a world, embody a character, and forget we are the dreamer. No splitting is required, only immersion.

Where Kastrup theorizes metaphysical trauma, CDIT sees symbolic play. Where panpsychism imagines innumerable flickers of proto-awareness, CDIT recognizes a single luminous field, lucid, dreaming, undivided.

Differentiation from Spinoza and Other Nondual Philosophies

Spinoza's metaphysics offers one of the most influential early articulations of a nondual worldview. He held that there is only one substance, God or Nature, and that everything else is a mode or expression of this indivisible substance (Spinoza, 1677/1985). For Spinoza, mind and body are two attributes of this same underlying reality, and each individual is a modification of the singular divine essence.

CDIT shares this nondual foundation but diverges in two important ways. First, CDIT avoids labeling awareness as "God" or attributing to it the moral, teleological, or structural connotations of divine nature. Awareness in CDIT is not omniscient, omnipotent, or moral, it is not a personified being but a natural presence, akin to gravity or electromagnetism. It expresses not from intention or necessity, but from innate potential.

Second, CDIT frames individuation not as a modification of a substance, but as a narrative rendering in a dream. Spinoza's modes still carry ontological weight as partial expressions of substance; CDIT sees dream characters not as parts of the dreamer, but as appearances arising within a single dreaming field. The dreamer doesn't become them, it dreams them.

Spinoza's system also leans on a rationalistic geometry of cause and substance, suggesting that everything must unfold from divine necessity. CDIT is more phenomenological: it starts from lived experience and the observation that awareness spontaneously renders symbolic worlds without an external cause. The dreamer is not fulfilling a logical mandate, it is playing with possibility.

Other Related Theories Briefly Addressed

Neutral Monism: This view posits a fundamental "neutral" stuff from which both mind and matter derive (James, Russell). But neutral monism still implies a substrate more basic than consciousness. CDIT argues that awareness is already the fundamental, not neutral, but active, expressive, and sufficient.

Cosmopsychism: Recent versions (e.g., Nagasawa & Wager, 2016) suggest the cosmos itself is conscious and that individual minds are aspects of that larger mind. CDIT agrees in part, the universe is within awareness but rejects the view that the cosmos is the subject. The cosmos is a dream scene, not the dreamer.

Whitehead's Process Philosophy: Whitehead emphasized becoming and experiential events as the basic units of reality. CDIT resonates with this dynamic ontology but disagrees that "actual occasions" form a metaphysical pluralism. CDIT holds that all apparent events arise within a single dreaming awareness, rather than being individually real.

Plato: In Plato's theory of Forms, ultimate reality consists of abstract, perfect ideals beyond physical matter. CDIT differs in that it does not posit an abstract, transcendent realm but a living, experiential one. Reality is not derived from eternal concepts but rendered dynamically within the dream of awareness.

Descartes: Descartes began with the certainty of "I think, therefore I am," grounding knowledge in the existence of a thinking self. CDIT starts even earlier, not with thought, but with the awareness in which thought arises. The self is not a thinking substance but a dreamed persona within awareness.

Schopenhauer: Schopenhauer asserted that the underlying reality is a blind, striving will, irrational and impersonal. CDIT rejects this drive-based metaphysics. Awareness is not a will or a need. It does not strive or lack. It simply dreams, not from desire, but from inherent symbolic unfolding.

CDIT is more parsimonious than these other models. It does not posit a hierarchy of substances, agents, or emergent processes that themselves require explanation. Whether it's Spinoza's impersonal "substance," cosmopsychism's universal mind, or process philosophy's cascading "actual occasions," each ultimately faces the problem of infinite regress. We are still left asking: Where did that come from?

Who or what caused the cause? What made the god, the field, or the fundamental process?

CDIT dissolves this regress not by adding another layer, but by stepping outside of the causal chain altogether. It proposes that awareness is not a thing within time, but the precondition for time and causality themselves. It was never created. It did not emerge from anything more fundamental. It is not a field in the physical sense, but rather the condition in which fields, minds, matter, and experience appear, much as a dream appears within the still presence of sleep.

This is not metaphysical speculation. It is the one feature of reality that can be confirmed directly. Between any two thoughts, prior to memory or perception, there is a silent presence, a knowing that cannot be grasped but never disappears. It is always now. We cannot get behind it, because it is what makes the very act of seeking, knowing, and experiencing possible.

CDIT requires no further explanation because it arrives at the only thing that does not need one. Awareness is not a creation; it is what makes creation, causation, and explanation intelligible. All other models still dream of something deeper. CDIT invites us to wake up to what is already most immediate.

The Lucid Dreamer Analogy

Skilled lucid dreamers have reported the ability to shift into different characters during a single dream (Voss et al., 2009; LaBerge, 1985). Rather than staying tethered to a single dream-ego, they can "jump" perspectives, becoming the observer, then another participant, then back again.

CDIT draws a parallel here: awareness is the ultimate lucid dreamer. Just as a dreamer can shift identities without truly being any of them, awareness can stream through any symbolic lens it renders.

Your current life feels continuous because this rendering is stabilized, but it is still a dream role, not a fixed identity. In this light, being "you" now is not a selection out of many, but simply the coherent perspective generated in this moment of the dream.

Form as Dream, Not Destiny

This same principle applies to the forms we inhabit, our bodies, minds, and histories are not predestined outcomes, nor are they chosen by a separate creator. They emerge from the same spontaneous dreaming capacity of awareness that we glimpse every night in personal dreams, but here without the filters of personal memory or ego. Human beings are not accidents, nor are they celestial masterpieces; we are natural symbolic unfoldings within the larger field of experience. Our existence, rich with language, love, wonder, and sorrow, is not the result of design; it is the inevitable flowering of awareness encountering itself through endless perspectives.

Recursive Dreaming and the Echo of Awareness

Reality may be a dream within which dreaming continues. Every layer adds nuance, each dream nested inside the next, not infinitely, but richly. This layering doesn't fragment the dreamer; it expands the reach of awareness into every depth and distortion.

CDIT views recursive dreaming as evidence of awareness's symbolic recursion, not mechanical, but expressive. The more perspectives the dreamer simulates, the more intricately it knows itself.

Final Thoughts

So how does awareness create a universe? Not by building it, not by designing it, and not by choosing it. It arises like a dream arises, spontaneously, consistently, coherently, from a timeless and formless field that is not nothing, but the possibility of everything.

This does not diminish the beauty of life. It deepens it. It means every leaf, every breath, every feeling of love or grief is not a byproduct of inert matter, but an event in the field of consciousness, a ripple in the sea of being.

CDIT shows that reality does not need a creator; it needs a dreamer. And the dreamer is you, not the character, but the timeless awareness behind all characters.

Chapter 20

Testing the Dream—Can CDIT Be Explored or Validated?

If this entire work is built on the premise that reality is akin to a dream, projected from and sustained by pure awareness, then the natural question becomes: can we test this?

Philosophical idealism is often dismissed as unfalsifiable. It's seen as metaphysics, not science. But the boundary between science and metaphysics is not always as firm as it appears. Many widely respected scientific theories, string theory, the multiverse, even aspects of quantum gravity, are not directly testable in the laboratory. What makes a theory valuable is not only its testability, but its explanatory power, coherence, and ability to unify disparate phenomena.

Still, the desire to probe a theory's implications is natural. If CDIT is more than poetic speculation, if it offers a serious model of consciousness and reality, it should yield consequences. It should open new lines of exploration: experimental, experiential, and philosophical.

Phenomenological Experiments

Start with dreams. Every human experiences a world generated from within, a coherent space, unfolding in time, shaped by memory, emotion, and thought. Lucid dreams go a step further: the dreamer becomes aware they are dreaming, and can interact with the content consciously (LaBerge, 1985). In meditation, sensory deprivation, or under psychedelics, similar phenomena occur: boundaries dissolve, identity loosens, and awareness becomes central (Griffiths et al., 2006; Wallace, 2007).

These states offer a kind of internal experiment. In some cases such as deep meditation, sensory deprivation, psychedelic states, people report that when thoughts, perceptions, and bodily sensations fall away, a witnessing presence remains. In other cases, such as dreamless sleep or anesthesia, there may be no remembered content, yet awareness reappears seamlessly with the return of experience. From the CDIT perspective, both patterns suggest that awareness is not generated by the brain, but is the field in which all content appears, even when that content is absent from memory.

Quantum Physics and the Role of the Observer

The observer problem in quantum mechanics continues to perplex. Why does the behavior of particles seem to depend on measurement? Why do certain experiments (like the delayed-choice quantum eraser) suggest that observation can influence the past (Wheeler, 1978; Kim et al., 2000)?

Idealism offers a powerful answer: because observation is not something that occurs in the world, it is the act of awareness itself that renders the world. Without awareness, there is no particle, no path, no outcome.

CDIT incorporates these findings not as proofs but as signs that reality behaves more like a symbolic rendering than a machine. It interprets the quantum world as dream-logic: potential becomes actual only when experienced, and what is experienced depends on the storyline being rendered.

Biological Anomalies and Distributed Intelligence

Michael Levin's work with bioelectric patterning reveals that form and regeneration are not fully explained by genes. Tissues "know" how to build and rebuild themselves using non-local, system-wide information fields (Levin, 2020).

CDIT suggests this model is not encoded in molecules but symbolically rendered within the dream. The body is not a machine reacting to stimuli, it is a story unfolding in form. Awareness, not atoms, determines identity and regeneration.

The Continuity of Awareness Across Brain States

Whether dreaming, awake, hallucinating, sedated, or near death, many report a continued sense of "beingness" that seems to persist beneath changing cognitive content (Greyson, 2000). Even when memory and self-narrative dissolve, this basic capacity for experience, awareness itself, is not destroyed, but reappears seamlessly with the return of content. From the CDIT perspective, apparent gaps such as dreamless sleep or anesthesia are not literal cessations of awareness, but intervals in which the dream narrative withholds intermediate frames. When the story resumes, it does so within the same underlying field. This constancy supports the view that the brain functions as an interface, not a generator; awareness is like the screen behind the movie, present regardless of whether a scene is currently playing.

Anomalous Phenomena at the Edges of Mind

Telepathy, precognition, synchronicity, and shared dream experiences are often dismissed as fringe. Yet they persist across cultures and contexts. Controlled Ganzfeld experiments in parapsychology have reported statistically significant evidence of telepathy-like effects (Bem & Honorton, 1994).

Within CDIT, these are not supernatural anomalies, they're consistent with a shared dream reality. Minds may not be separate hard drives but localizations within a single awareness. Synchronicity and intuitive resonance may reflect coherence across perspectives inside the dream.

Simplicity and Philosophical Economy

Materialism posits non-aware matter giving rise to experience, yet experience is the only thing we know directly (Chalmers, 1996). Dualism posits two substances, mind and matter, without explaining how they interact. Both frameworks rely on something unseen to explain what is seen.

CDIT needs only awareness. Awareness can produce form (as in dreams), causality (as in stories), and diversity (as in multiple perspectives). It is the most parsimonious model: one principle, experienced directly, sufficient to explain all else.

From this perspective, CDIT does not anticipate future discoveries in the form of new fundamental substances or hidden mechanisms. Rather, it anticipates a continued shift toward constraint-based explanations. Just as gravity has been reinterpreted as geometry rather than force, and dark matter and dark energy function as inferred requirements for large-scale coherence rather than observed entities, future advances may increasingly take the form of limits, consistency conditions, and structural constraints. These would not act on matter as causes, but define the conditions under which a shared, stable world of experience can appear at all.

A parallel pattern is already visible in biology. Levin's research in morphogenesis has shown that genetic and chemical processes alone do not fully determine biological form. Instead, global bioelectric patterns act as constraint fields that guide how cells organize, repair, and regenerate into coherent body plans. These fields do not push cells mechanically or encode construction instructions. They function as boundary conditions that define which anatomical configurations are stable, permitted, or corrected when disrupted.

In this sense, CDIT does not predict a new force acting on matter, but it does anticipate that physics and biology alike will continue to formalize global consistency constraints that are already implicitly required for reality to remain shared, stable, and free of experiential contradiction.

Rather than competing with scientific discovery, CDIT offers a lens through which the trajectory of discovery itself becomes intelligible, as an ongoing shift away from hidden substances and toward the rules that make coherent experience possible.

A Bridge to the Philosophical Core

None of these points "prove" CDIT in the scientific sense. But they point toward it, support its coherence, and invite serious inquiry. CDIT is not an empirical theory, but an experiential model. It asks not for belief, but for recognition. It does not claim to describe reality as an object; it asks you to recognize yourself as the subject, the field in which experience arises, regardless of whether content is continuously present.

The question is no longer whether CDIT can be tested in a lab. The question is whether you can see the dream while still inside it.

Still, concerns remain:

Is it too poetic? Too subjective? Too immune to critique?

These are real questions, and they deserve real answers.

In the next chapter, we meet those critiques head-on.

Chapter 21

In Defense of CDIT

The Challenge of Defending a Dream

No paradigm shift escapes critique. Cognitive Dream Interface Theory (CDIT) will, like all serious metaphysical models, attract questions from skeptics. Some will challenge its scientific basis, others its philosophical coherence. Some will mistake it for solipsism. Others will dismiss it as poetic metaphor untethered from empirical rigor.

But CDIT does not rest on belief. It rests on the one thing no theory can deny: direct experience. It offers not only a model of reality, but a mirror through which reality begins to recognize itself.

Is CDIT Falsifiable?

A common critique is that CDIT, like many metaphysical frameworks, is not falsifiable in the Popperian sense. It cannot be tested with a single experiment and declared true or false. But this critique misunderstands CDIT's domain. It is not a scientific theory competing with physics. It is a metaphysical model explaining why science itself appears the way it does.

CDIT does not replace empirical inquiry, it explains it from within. Physics, chemistry, biology, and neuroscience all arise as consistent renderings within the dream interface of awareness.

Comparison to Simulation and Theistic Models

Today, two popular metaphysical models dominate imagination: the simulation hypothesis and theistic creation. Both posit an external entity, a programmer, a god, who brings reality into being. But both suffer from the same flaw: dualism. They require something outside the system to create the system.

Who built the simulator? Who created the god? These questions trigger an infinite regress, because the creator itself would need explaining.

CDIT avoids this trap entirely. It does not posit a being that came into existence at some point in time. Awareness, in CDIT, is not an object within the system, but the timeless condition that allows any system to appear. It is not "made" of anything else and does not arise from prior causes. Asking "Who created awareness?" is like asking "What is north of the North Pole?" The question mistakes the nature of what's being described.

Unlike a god or a simulator, both of which remain hypothetical and hidden, awareness is directly available to anyone willing to look. It is the one thing you do not need faith to verify; it is present before any belief, opinion, or theory. Meditation and contemplative practice can make its presence more vivid, but it requires no leap of trust in the unseen. Awareness is self-evident: the undeniable ground in which every perception, thought, and feeling appears.

There is no need for a designer when awareness can dream. No need for external machinery when symbolic interface can arise naturally, just as it does nightly in every human mind.

Science as Support but Not Requirement

Quantum entanglement, temporal non-locality, the block universe, evolutionary interface theory, all point toward a reality less objective and material than classical science imagined (Hoffman, 2019; Kim et al., 2000).

CDIT embraces scientific findings but does not depend on them. Awareness is self-evident, prior to any instrument, observation, or theory. Like pain or color, it must be recognized, not proven.

You do not need belief to be conscious. You are conscious. You do not need consensus to experience awareness. It is already here.

Dreams as Direct Demonstrations

Skeptics often demand proof that mind can generate coherent realities. But we encounter such proof every night.

Dreams demonstrate that a mind can create immersive, original, richly textured worlds without planning or memory (LaBerge, 1985). People, landscapes, histories, emotions are spontaneously rendered, internally consistent, and emotionally powerful.

CDIT holds that waking reality is dream-reality with higher narrative coherence. The waking world feels more stable, but both waking and dreaming are symbolic renderings within the same stable awareness.

Gödel's Incompleteness and the Limits of Objectivity

Kurt Gödel's incompleteness theorems demonstrated that any sufficiently rich formal system, specifically, one capable of expressing arithmetic, contains true statements that cannot be proven from within the system itself (Gödel, 1931). While Gödel was writing about mathematical logic, his work has been interpreted more broadly by philosophers and cognitive scientists to highlight the limitations of self-referential systems.

In this spirit, CDIT applies Gödel's insight metaphorically: materialism and simulation theory are "closed systems" that attempt to explain consciousness using the symbolic constructs that consciousness itself renders. Like reading the script of a play and trying to infer the stage it was performed on, materialist and simulation-based models try to reconstruct the source of experience from within experience itself. But the script doesn't contain the stage, it assumes it. These models loop endlessly within the logic they generate, never stepping outside the system to explain how the system itself appears.

CDIT avoids this recursive trap. It does not try to explain the dream from within the dream. Instead, it recognizes awareness as the open canvas on which all symbols are painted, not a product of the rendering, but its precondition. Awareness is not emergent from complexity or matter. It is the silent background that allows all emergence to appear.

The Illusion of Emergence: Grains, Molecules, and Minds

Materialism assumes that complex phenomena emerge from the accumulation and arrangement of parts; that *quantity becomes quality*. But this model falters upon close inspection. How many grains of sand are required to make an anthill? At what exact molecular count does water become "wet"? These are not thresholds that can be precisely defined. Rather, they are **appearances of coherence**, gestalt renderings shaped by context.

CDIT applies the same principle to consciousness. There is no precise number of neurons or configuration of circuits that gives rise to awareness. Like the anthill or the liquidity of water, consciousness is not something that "emerges" from complexity. It is the presence in which such symbolic constructs appear. Awareness is not built from parts; it is what allows parts to appear in the first place.

Brain scans, despite their growing sophistication, have yet to isolate a signature of pure awareness. What they detect are patterns of activity *within* the interface, shifting correlates of thought, emotion, and perception, not the presence in which those patterns arise.

CDIT reframes emergence not as a process but as a symbolic backfill. Once something appears: an anthill, a personality, a moment of consciousness, the dream generates a plausible history. But the origin itself is never found. This is akin to the uncertainty principle in quantum physics: just as one cannot simultaneously know a particle's position and momentum with precision, one cannot locate both the appearance and the exact point of its emergence. To perceive something as present is to obscure the process of its becoming.

And just as we don't question how molecules conspired to create "wetness," the dream of selfhood fills in its backstory only after it has begun. You find yourself in the middle, already experiencing, and only then do memory, biology, and identity begin to narrate how you got there. In CDIT, emergence is a narrative artifact. There is no causal chain behind awareness, only a coherent rendering within it.

The Paradox of Time and the Vanishing Past

CDIT also reframes the illusion of time through the lens of symbolic rendering. In both waking life and dreams, experience seems to unfold moment by moment, with a past receding behind us and a future rushing forward. But upon inspection, each "past" event is not a preserved object we move away from, it is part of the current rendering. As with dreams, the storyline appears continuous but is actually being stitched into the present. The words you read, the thoughts you recall, the actions you believe you took yesterday, none are retrieved from a real timeline. They are symbolically rendered now to preserve coherence. In this way, the past is not a place we came from, but a narrative texture added to the moment. Awareness never actually moves through time; it simply witnesses each immersive frame as it appears. Like trying to find the exact grain of sand where an anthill began, the attempt to locate when "now" becomes "then" collapses under scrutiny. There is only this: the present rendering, and the timeless witness aware of it.

Symmetry as Evidence of Dream Coherence

Across the universe, we encounter a striking recurrence of symmetry. Spiraling galaxies resemble neural networks, fractal branching appears in trees and vascular systems, and the bilateral structure of living bodies reflects a deep-seated pattern in nature. Even in mathematics, the Fibonacci sequence arises repeatedly in flower petals, pinecones, hurricanes, and galaxies. These forms are not only beautiful but functionally embedded in physics, biology, and chemistry. In physical law, symmetry is the basis for conservation principles. In biology, fractal geometry supports efficient growth and distribution. In chemistry, molecular structures often display repeating geometries that support stability and reactivity.

Within CDIT, such symmetry is not the result of external design or divine intention, but rather a reflection of the coherence inherent to the dream itself. Just as nightly dreams express recurring themes, images, and emotional rhythms, the waking world also unfolds as a symbolic rendering generated by awareness. These patterns do not arise from conscious calculation but emerge naturally from awareness's intrinsic tendency toward ordered expression.

The presence of symmetry, rhythm, and proportion is not evidence of a creator, but of a dreaming process that privileges coherence. The Fibonacci spiral is not imposed on nature from outside but unfolds as a natural expression of the same awareness that dreams time, space, and form. In this view, symmetry is not merely a byproduct of physical law, it is a symbolic signature. It signals that what we take to be objective structure is, at its root, the output of a symbolic rendering process within awareness whose nature is to express coherent experience.

Common Critiques and CDIT's Response

1. "Isn't CDIT just solipsism?"

No. Solipsism claims that only your individual mind exists. CDIT proposes that there is one shared awareness dreaming many localized perspectives. Your character is not the only character; all dream characters arise in the same awareness, though each experiences only its localized narrative.

There are many windows, but one sky.

2. "But brain damage changes consciousness"

While it may affect brain function, it will not impact awareness. This is expected if the brain is the interface, not the source. Damaging a TV antenna scrambles the signal; it does not prove the signal originates in the antenna. Similarly, altering the brain alters the rendering of experience without disproving the existence of awareness itself.

CDIT frames the brain as an interface modulating experience, not generating it.

3. "Isn't CDIT just subjective idealism?"

No. CDIT asserts that objectivity, the shared consistency of space, time, and form, emerges as a stabilized symbolic rendering. The dream obeys internal rules to make shared experience possible. Subjectivity and objectivity are both modes within awareness.

The "external world" is real, but it is real as rendered experience, not as an independent material substratum.

Awareness as the Only Necessary Assumption

Why does awareness dream? CDIT reframes the question: it dreams because it can. Awareness does not lack, seek, or strategize, it expresses.

Just as gravity pulls, electromagnetism radiates, or the strong nuclear force binds, awareness unfolds as experience. It is not anthropomorphic, purposive, or psychological in any human sense; it is a naturally occurring presence whose intrinsic property is the rendering of symbolic form. We readily accept that evolution can be a naturally driving force, shaping life from simple beginnings into complex forms, and that elemental particles can forge seemingly purposeful destinies without any conscious design. Why, then, should it be so difficult to accept that the base layer of reality, awareness itself, could naturally "evolve" its own diversity of experience by dreaming, conjuring imagery, and generating worlds? In this sense, the spontaneous variety of awareness's renderings is its form of evolution: a continual exploration of its own possibilities.

CDIT proposes that awareness renders experience not to complete itself, but to explore itself. This spontaneous unfolding needs no external justification. There is no "why" behind awareness, just as there is no "why" behind the fact that mass curves spacetime. Awareness is the presence, and dreaming is how that presence expresses itself.

Another common critique is that CDIT might imply an infinite chain of dreamers dreaming dreamers. But this misunderstands the model. CDIT holds that there is only one awareness, unchanged and indivisible, through which all perspectives appear. Dream layering is symbolic, not structural. The same awareness is present in waking, dreaming, and deep sleep, and never changes. Infinite regress requires multiple agents in time, but awareness is outside of time. It does not dream other dreamers; it dreams the illusion of many from within itself. The coherence of waking life suggests it is the primary rendered dream, but all content, vivid or fractured, appears in the same awareness, which is the only thing we ever truly know is real.

The Strength of the Model

CDIT does not explain everything. It does not offer predictive equations or technological shortcuts. What it offers is coherence across:

- Why time feels linear but may not be
- Why perception is functional but not truthful
- Why consciousness appears local but is nonlocal
- Why matter seems objective but is experienced subjectively

CDIT unifies these under one principle: awareness itself, direct, accessible, and requiring no new entities or hidden variables.

Awareness does not come from anywhere, because "coming from" is a concept within the dream. Origin stories imply time, causality, and separation, all of which are part of the rendering that awareness generates. Awareness is not within the picture; it is the condition that makes the picture possible. Like silence underlying music, it is always there, even when covered by sound.

And unlike other theories that propose gods, fields, or minds that require further justification, CDIT points to what is directly available now: the undeniable fact that something is happening. At the root of every experience is the knowing of it: awareness. Everything else is content; awareness is the container that cannot itself be contained. There is nothing more foundational, because there is nothing else that does not depend on it.

Final Thoughts: Toward Rigorous Recognition

Many resist CDIT because we are taught that consciousness must require a brain, neurons, or complexity. Yet neuroscience has no mechanism for how subjective experience arises from neural activity. The brain is measurable; awareness is not. This absence is no mystery in CDIT: the brain is part of the dream, a rendered symbol of localized perspective. Awareness is not represented in the brain for the same reason the movie screen is not visible inside a scene, it is not an object among others, but the condition in which all objects appear.

We accept gravity and the other fundamental forces as naturally occurring, without machinery or intention. We do not demand an origin story for why mass curves spacetime. Yet with awareness, we ask where it came from or what preceded it. CDIT treats awareness the same way: as a foundational feature of nature, not created or caused, and unlike those forces, it is in direct contact with us.

When sensory input is reduced, in sleep, under psychedelics, or in sensory deprivation, spontaneous worlds still appear.

Minds, when unstimulated, generate content. Just as apples fall, awareness dreams. Experience is not a side effect of complexity; it is the expression of a naturally occurring phenomenon whose nature is to render evolving experience.

Skepticism is healthy. Inquiry is necessary. CDIT welcomes both. But the ultimate test is direct: look not outward, but inward. See whether there is anything more fundamental than the awareness in which the world appears. Every thought, doubt, and argument happens within the dream. Awareness is the one phenomenon that requires no external validation, because it is the ground of validation itself. It was never part of the dream. It is the dreamer.

Epilogue

Gently Down the Stream

There is a children's song that almost everyone knows:

Row, row, row your boat
Gently down the stream.
Merrily, merrily, merrily, merrily,
Life is but a dream.

We learn it before we understand metaphor, before we grasp the mystery of being, before we ask the deeper questions of where we came from or what this all means. And yet, in its simplicity, it contains a profound truth, one that the greatest philosophers, physicists, and sages have only hinted at in more complicated terms.

Row your boat: take action.

Gently: without force.

Down the stream: follow the current of awareness, not against it.

Merrily: with joy.

And finally: **life is but a dream.**

In this book, we've explored the possibility that everything we take to be real, our bodies, our minds, our planet, our universe, may be dreamlike in nature. Not unreal as in imaginary, but dreamed from within a deeper field of awareness, the one certainty no model of reality can deny.

The framework presented here, **Cognitive Dream Interface Theory (CDIT)**, is the result of an independent exploration across fields as varied as neuroscience, quantum physics, and philosophical idealism. It builds upon many prior ideas, but its synthesis is original: an attempt to offer a coherent, naturalistic view of existence as a dream-like interface rendered within awareness. It is not offered as the final word, but as an invitation to see the nature of reality, and yourself, in a new light.

We've seen how this perspective aligns with the puzzles of quantum physics, the gaps in materialist neuroscience, and the timeless intuitions of idealist philosophy. We've argued that awareness is not a byproduct of brain chemistry or an emergent property of evolution, but the foundation of all experience. And we've shown how this dream-like model can account for the continuity of consciousness, the illusion of time, the nature of death, and the profound interconnectedness of all beings.

This is not just philosophy. **It is a way of being.**

To live as if life is a dream, your dream, is to carry the world with a lighter hand. You stop grasping. You stop resisting. You start observing. You start engaging with the mystery, not as something to conquer, but as something to wonder at.

This is not detachment or nihilism. It is clarity.

It means recognizing that pain, joy, struggle, and triumph are all part of a story unfolding on the canvas of awareness, and that you are not merely in the story. **You are the field in which the story happens.**

The Role You Were Always Meant to Play

In the dream of awareness, the story of "you" arises with the moment, not by accident, not by assignment, but as the natural coherence of this scene. As in nightly dreams, you appear at the center of your own experience, because that is how immersion works. The history you remember, the body you inhabit, and the world you see are part of the scaffolding that makes this vantage point feel inevitable.

In the shared dream, there is no higher or lower, no self more real than another. Each perspective is a different window for the dreamer to see itself. Through this window, awareness lives your life, and through every other window, it lives all the rest.

Letting Go of Old Maps

Letting go of materialism is not easy. It is as radical as accepting that the earth is not flat, or that what appears as human evolution is part of a larger unfolding, rather than the ultimate cause of our existence. Paradigm shifts are always unsettling at first.

But the dream model, the CDIT framework, offers a view that does not require blind faith, advanced degrees, or theoretical equations. It offers something simpler, more direct, and ultimately more profound: the recognition of your own awareness, here and now.

You do not need to study quantum mechanics to realize you are dreaming. You only need to notice the dream happening.

This shift is not a retreat from science. It is science expanded, turned inward toward the one phenomenon that cannot be measured, yet without which nothing could be measured at all.

And in that recognition, the old maps fall away.

Reality is not inert matter.

It is the living dream of awareness itself.

And as it turns out, the children were right all along:

Life is but a dream.

Author's Note

This book began as a question, not a spiritual one, but a scientific one: **What is the most plausible explanation for the fact of conscious awareness?**

As a physician trained in biology and immersed in the materialist worldview, I never expected to land where I did. For years, I accepted without question the prevailing story: that matter gave rise to mind through evolutionary complexity. That consciousness was a byproduct, a fluke of chemistry and chance.

But the deeper I looked, the more that foundation cracked. Neuroscience could map brain activity, but not experience. Physics could describe particles, but not presence. Evolution could explain the adaptations of bodies, but not the fact that anything is aware in the first place.

The more carefully I examined the nature of awareness itself, the more implausible it seemed that matter alone could account for it. Awareness was not something I had, it was what I was. And from that realization, everything else followed.

This is not a call to belief. It is an invitation to notice.

To see that space, time, matter, and even the story of evolution may be parts of a grand, coherent rendering within awareness, **not the cause of awareness**.

I don't claim to have all the answers. No single book, model, or theory can fully capture the mystery of existence.

But if this framework resonates, I encourage you not to take my word for it.

Instead:

Look for the one who is looking.

References

References for Chapter 1

Barbour, J. (1999). *The End of Time: The Next Revolution in Physics*. Oxford University Press.

Bell, J. S. (1964). On the Einstein-Podolsky-Rosen paradox. *Physics Physique Физика*, 1(3), 195–200.

Hoffman, D. D. (2019). *The Case Against Reality: Why Evolution Hid the Truth from Our Eyes*. W. W. Norton & Company.

Wheeler, J. A. (1990). Information, physics, quantum: The search for links. In W. H. Zurek (Ed.), *Complexity, Entropy, and the Physics of Information* (pp. 3–28). Addison-Wesley.

References for Chapter 2

Adlam, E. (2014). The problem of confirmation in the Everett interpretation. *Studies in History and Philosophy of Modern Physics*, 47, 21–32.

Bell, J. S. (1964). On the Einstein-Podolsky-Rosen paradox. *Physics Physique Физика*, 1(3), 195–200.

Chalmers, D. J. (1996). *The Conscious Mind: In Search of a Fundamental Theory*. Oxford University Press.

Hoffman, D. D. (2019). *The Case Against Reality: Why Evolution Hid the Truth from Our Eyes*. W. W. Norton & Company.

Wheeler, J. A. (1990). Information, physics, quantum: The search for links. In W. H. Zurek (Ed.), *Complexity, Entropy, and the Physics of Information* (pp. 3–28). Addison-Wesley.

References for Chapter 3

Alcaro, A., Carta, S., & Panksepp, J. (2017). The affective core of the self: A neuro-archetypical perspective on the foundations of human (and animal) subjectivity. *Frontiers in Psychology*, 8, 1424. https://doi.org/10.3389/fpsyg.2017.01424

Brewer, J. A., Garrison, K. A., & Whitfield-Gabrieli, S. (2013). What about the "self" is processed in the posterior cingulate cortex? *Frontiers in Human Neuroscience*, 7, 647. https://doi.org/10.3389/fnhum.2013.00647

Damasio, A. (1999). *The Feeling of What Happens: Body and Emotion in the Making of Consciousness*. Harcourt.

Gallagher, S. (2000). Philosophical conceptions of the self: Implications for cognitive science. *Trends in Cognitive Sciences*, 4(1), 14–21.

Metzinger, T. (2009). *The Ego Tunnel: The Science of the Mind and the Myth of the Self*. Basic Books.

Seth, A. K. (2021). *Being You: A New Science of Consciousness*. Faber & Faber.

References for Chapter 4

Carhart-Harris, R. L., & Friston, K. J. (2019). REBUS and the anarchic brain: Toward a unified model of the brain action of psychedelics. *Pharmacological Reviews, 71*(3), 316–344.

Chalmers, D. J. (1995). Facing up to the problem of consciousness. *Journal of Consciousness Studies, 2*(3), 200–219.

Engel, A. K., & Singer, W. (2001). Temporal binding and the neural correlates of sensory awareness. *Trends in Cognitive Sciences, 5*(1), 16–25.

Gazzaniga, M. S. (2000). Cerebral specialization and interhemispheric communication. *Brain, 123*(7), 1293–1326.

Hoffman, D. D. (2019). *The Case Against Reality: Why Evolution Hid the Truth from Our Eyes*. W. W. Norton & Company.

Levin, M. (2021). Morphological computation and developmental bioelectricity: Rethinking the relationship between the brain and behavior. *Philosophical Transactions of the Royal Society B: Biological Sciences, 376*(1821), 20190752.

Metzinger, T. (2009). *The Ego Tunnel: The Science of the Mind and the Myth of the Self*. Basic Books.

Treisman, A. (1996). The binding problem. *Current Opinion in Neurobiology, 6*(2), 171–178.

References for Chapter 5

Adlam, E. (2014). The problem of confirmation in the Everett interpretation. *Studies in History and Philosophy of Modern Physics*, 47, 21–32.

Bell, J. S. (1964). On the Einstein-Podolsky-Rosen paradox. *Physics Physique Физика*, 1(3), 195–200.

Carroll, S. (2019). *Something Deeply Hidden: Quantum Worlds and the Emergence of Spacetime*. Dutton.

Hoffman, D. D. (2019). *The Case Against Reality: Why Evolution Hid the Truth from Our Eyes*. W. W. Norton & Company.

Susskind, L. (1995). The world as a hologram. *Journal of Mathematical Physics*, 36(11), 6377–6396.

Wigner, E. (1967). Remarks on the mind-body question. In *Symmetries and Reflections* (pp. 171–184). Indiana University Press.

References for Chapter 6

Avant, L. L. (1965). Vision in the Ganzfeld. *Psychological Bulletin*, 64(4), 246–258.

Heron, W. (1957). The pathology of boredom. *Scientific American*, 196(1), 52–56.

Merabet, L. B., et al. (2004). Visual hallucinations during prolonged blindfolding in sighted subjects. *Nature Neuroscience*, 7(6), 625–626.

Siclari, F., Baird, B., Perogamvros, L., et al. (2017). The neural correlates of dreaming. *Nature Neuroscience*, 20(6), 872–878.

Zubek, J. P. (1969). *Sensory Deprivation: Fifteen Years of Research*. Appleton-Century-Crofts.

References for Chapter 7

Adlam, E. (2014). The problem of confirmation in the Everett interpretation. *Studies in History and Philosophy of Modern Physics*, 47, 21–32.

Barbour, J. (1999). *The End of Time: The Next Revolution in Physics*. Oxford University Press.

Bell, J. S. (1964). On the Einstein-Podolsky-Rosen paradox. *Physics Physique Физика*, 1(3), 195–200.

Einstein, A. (1916). The foundation of the general theory of relativity. *Annalen der Physik*, 49(7), 769–822.

Rovelli, C. (2018). *The Order of Time*. Riverhead Books.

Siclari, F., Baird, B., Perogamvros, L., et al. (2017). The neural correlates of dreaming. *Nature Neuroscience*, 20(6), 872–878.

Tegmark, M. (2014). *Our Mathematical Universe: My Quest for the Ultimate Nature of Reality*. Alfred A. Knopf.

Wigner, E. (1967). Remarks on the mind-body question. In *Symmetries and Reflections* (pp. 171–184). Indiana University Press.

References for Chapter 8

Adlam, E. (2014). The problem of confirmation in the Everett interpretation. *Studies in History and Philosophy of Modern Physics*, 47, 21–32.

Griffiths, R. R., Richards, W. A., McCann, U., & Jesse, R. (2006). Psilocybin can occasion mystical-type experiences having substantial and sustained personal meaning and spiritual significance. *Psychopharmacology*, 187(3), 268–283.

Hoffman, D. D. (2019). *The Case Against Reality: Why Evolution Hid the Truth from Our Eyes*. W. W. Norton & Company.

Levin, M. (2021). Reprogramming cells and tissues: Building a biological mind from the ground up. *Cell*, 184(7), 1915–1929.

Stern, D. N. (1985). *The Interpersonal World of the Infant: A View from Psychoanalysis and Developmental Psychology*. Basic Books.

References for Chapter 9

Barbour, J. (1999). *The End of Time: The Next Revolution in Physics*. Oxford University Press.

Carhart-Harris, R. L., Leech, R., Williams, T. M., Erritzoe, D., Abbasi, N., Bargiotas, T., ... & Nutt, D. J. (2014). The entropic brain: a theory of conscious states informed by neuroimaging research with psychedelic drugs. *Frontiers in Human Neuroscience*, 8, 20.

Feinstein, J. S., Khalsa, S. S., Yeh, H., Wohlrab, C., Simmons, W. K., Stein, M. B., & Paulus, M. P. (2018). Examining the short-term anxiolytic and antidepressant effect of floatation-REST. *PLOS ONE*, 13(2), e0190292.

Gackenbach, J. (1991). *Conscious Mind, Sleeping Brain: Perspectives on Lucid Dreaming*. Springer.

Gallagher, S. (2000). Philosophical conceptions of the self: Implications for cognitive science. *Trends in Cognitive Sciences*, 4(1), 14–21.

Griffiths, R. R., Richards, W. A., McCann, U., & Jesse, R. (2006). Psilocybin can occasion mystical-type experiences having substantial and sustained personal meaning and spiritual significance. *Psychopharmacology*, 187(3), 268–283.

Josipovic, Z. (2010). Duality and nonduality in meditation research. *Consciousness and Cognition*, 19(4), 1119–1121.

LaBerge, S. (1985). *Lucid Dreaming*. Ballantine Books.

Nour, M. M., Evans, L., Nutt, D., & Carhart-Harris, R. L. (2016). Ego-dissolution and psychedelics: Validation of the Ego-Dissolution Inventory (EDI). *Frontiers in Human Neuroscience*, 10, 269.

Rochat, P. (2003). Five levels of self-awareness as they unfold early in life. *Consciousness and Cognition*, 12(4), 717–731.

Seth, A. (2021). *Being You: A New Science of Consciousness*. Faber & Faber.

Stumbrys, T., Erlacher, D., Schädlich, M., & Schredl, M. (2012). Induction of lucid dreams: A systematic review of evidence. *Consciousness and Cognition*, 21(3), 1456–1475.

Weiskrantz, L. (1997). *Consciousness Lost and Found: A Neuropsychological Exploration*. Oxford University Press.

Zeilinger, A. (2010). *Dance of the Photons: From Einstein to Quantum Teleportation*. Farrar, Straus and Giroux.

References for Chapter 10

Adlam, E. (2022). The reality of time symmetry. *Foundations of Physics*, 52(5), 1–38

Barbour, J. (1999). *The End of Time: The Next Revolution in Physics*. Oxford University Press.

Rovelli, C. (2018). *The Order of Time*. Riverhead Books.

Sacks, O. (1985). *The Man Who Mistook His Wife for a Hat and Other Clinical Tales*. Summit Books.

Schacter, D. L. (1999). The seven sins of memory: Insights from psychology and cognitive neuroscience. *American Psychologist*, 54(3), 182–203.

Squire, L. R. (2009). Memory and brain systems: 1969–2009. *The Journal of Neuroscience*, 29(41), 12711–12716.

References for Chapter 11

Carhart-Harris, R. L., Erritzoe, D., Williams, T., Stone, J. M., Reed, L. J., Colasanti, A., ... & Nutt, D. J. (2014). Neural correlates of the LSD experience revealed by multimodal neuroimaging. *Proceedings of the National Academy of Sciences*, 113(17), 4853–4858.

Gazzaniga, M. S. (2005). Forty-five years of split-brain research and still going strong. *Nature Reviews Neuroscience*, 6(8), 653–659.

Josipovic, Z. (2010). Duality and nonduality in meditation research. *Consciousness and Cognition*, 19(4), 1119–1121.

Searle, J. R. (1980). Minds, brains, and programs. *Behavioral and Brain Sciences*, 3(3), 417–457.

Turing, A. M. (1950). Computing machinery and intelligence. *Mind*, 59(236), 433–460.

Vyazovskiy, V. V. (2015). Sleep, recovery, and metaregulation: Explaining the benefits of sleep. *Nature Reviews Neuroscience,* 16(8), 479–480.

References for Chapter 12

Carhart-Harris, R. L., et al. (2014). Neural correlates of the LSD experience revealed by multimodal neuroimaging. *Proceedings of the National Academy of Sciences,* 113(17), 4853–4858.

Chalmers, D. J. (1996). *The Conscious Mind: In Search of a Fundamental Theory.* Oxford University Press.

Levin, M. (2021). Morphogenesis and the bioelectric code: The missing link between genes and anatomy? *Journal of Experimental Biology,* 224(1), jeb232047.

Merabet, L. B., Pascual-Leone, A., et al. (2004). Rapid and reversible recruitment of early visual cortex for touch. *Nature Neuroscience,* 7(12), 1196–1202.

Nagel, T. (1974). What is it like to be a bat? *The Philosophical Review,* 83(4), 435–450.

Suedfeld, P., & Steel, G. D. (2000). The environmental psychology of capsule habitats. *Annual Review of Psychology,* 51, 227–253.

References for Chapter 13

Bauer, P. J. (2007). *Remembering the Times of Our Lives: Memory in Infancy and Beyond.* Lawrence Erlbaum Associates.

Brewer, J. A., et al. (2011). Meditation experience is associated with differences in default mode network activity and connectivity. *Proceedings of the National Academy of Sciences,* 108(50), 20254–20259.

Carhart-Harris, R. L., Erritzoe, D., Williams, T., Stone, J. M., Reed, L. J., Colasanti, A., ... & Nutt, D. J. (2012). Neural correlates of the psychedelic state as determined by fMRI studies with psilocybin. *Proceedings of the National Academy of Sciences,* 109(6), 2138–2143.

Dennett, D. C. (1991). *Consciousness Explained*. Little, Brown and Company.

Feinstein, J. S., Khalsa, S. S., Yeh, H., Wohlrab, C., Simmons, W. K., Stein, M. B., & Paulus, M. P. (2018). Examining the short-term anxiolytic and antidepressant effect of Floatation-REST. *PLOS ONE*, 13(2), e0190292.

Nelson, K. (1996). *Language in Cognitive Development: Emergence of the Mediated Mind*. Cambridge University Press.

Whorf, B. L. (1956). *Language, Thought, and Reality: Selected Writings of Benjamin Lee Whorf*. MIT Press.

References for Chapter 14

Andrews-Hanna, J. R., Reidler, J. S., Sepulcre, J., Poulin, R., & Buckner, R. L. (2010). Functional-anatomic fractionation of the brain's default network. *Neuron*, 65(4), 550–562.

Barrett, L. F. (2017). *How Emotions Are Made: The Secret Life of the Brain*. Houghton Mifflin Harcourt.

Libet, B. (1985). Unconscious cerebral initiative and the role of conscious will in voluntary action. *Behavioral and Brain Sciences*, 8(4), 529–566.

Metzinger, T. (2003). *Being No One: The Self-Model Theory of Subjectivity*. MIT Press.

References for Chapter 15

Andrews-Hanna, J. R., Reidler, J. S., Sepulcre, J., Poulin, R., & Buckner, R. L. (2010). *Functional-anatomic fractionation of the brain's default network*. Neuron, 65(4), 550–562.

Barrett, L. F. (2017). *How Emotions Are Made: The Secret Life of the Brain*. Houghton Mifflin Harcourt.

Libet, B. (1985). *Unconscious cerebral initiative and the role of conscious will in voluntary action*. Behavioral and Brain Sciences, 8(4), 529–566.

Metzinger, T. (2003). *Being No One: The Self-Model Theory of Subjectivity*. MIT Press.

References for Chapter 16

Adlam, E. (2022). The problem of confirmation in the Everett interpretation. *Studies in History and Philosophy of Modern Physics*, 78, 11–22.

Bell, J. S. (1964). On the Einstein-Podolsky-Rosen paradox. *Physics Physique Физика*, 1(3), 195–200.

Borjigin, J., Lee, U., Liu, T., Pal, D., Huff, S., Klarr, D., ... & Mashour, G. A. (2013). Surge of neurophysiological coherence and connectivity in the dying brain. *Proceedings of the National Academy of Sciences*, 110(35), 14432–14437.

Levin, M. (2021). Morphogenesis as a model for programming the collective intelligence of cells and tissues. *Philosophical Transactions of the Royal Society B: Biological Sciences*, 376(1821), 20190766.

Maharaj, N. (1973/1990). *I Am That: Talks with Sri Nisargadatta Maharaj*. Translated by M. Frydman. Acorn Press.

Parnia, S., Spearpoint, K., & Fenwick, P. (2014). Near-death experiences, cognitive function, and psychological outcomes: A prospective study. *Resuscitation*, 85(12), 1799–1805.

Wheeler, J. A. (1978). The "past" and the "delayed-choice" double-slit experiment. In A. R. Marlow (Ed.), *Mathematical Foundations of Quantum Theory* (pp. 9–48). Academic Press.

References for Chapter 17

Aspect, A., Dalibard, J., & Roger, G. (1982). Experimental test of Bell's inequalities using time-varying analyzers. *Physical Review Letters*, 49(25), 1804–1807.

Einstein, A. (1915). The field equations of gravitation. *Sitzungsberichte der Preussischen Akademie der Wissenschaften zu Berlin*.

't Hooft, G. (1993). Dimensional reduction in quantum gravity. *arXiv:gr-qc/9310026*.

Susskind, L. (1995). The world as a hologram. *Journal of Mathematical Physics*, 36(11), 6377–6396.

Moser, E. I., Kropff, E., & Moser, M.-B. (2008). Place cells, grid cells, and the brain's spatial representation system. *Annual Review of Neuroscience*, 31, 69–89.

O'Keefe, J., & Nadel, L. (1978). *The Hippocampus as a Cognitive Map*. Oxford University Press.

Travis, F., & Shear, J. (2010). Focused attention, open monitoring and automatic self-transcending: Categories to organize meditative practices. *Consciousness and Cognition*, 19(4), 1110–1118.

Van Raamsdonk, M. (2010). Building up spacetime with quantum entanglement. *General Relativity and Gravitation*, 42(10), 2323–2329.

References for Chapter 18

Decety, J., & Jackson, P. L. (2004). The functional architecture of human empathy. *Behavioral and Cognitive Neuroscience Reviews*, 3(2), 71–100.

Greyson, B. (2000). Near-death experiences in a psychiatric outpatient clinic population. *Psychiatry Research*, 97(2–3), 161–167.

Strawson, G. (2006). Realistic monism: Why physicalism entails panpsychism. *Journal of Consciousness Studies*, 13(10–11), 3–31.

Wattles, J. (1996). *The Golden Rule*. Oxford University Press.

References for Chapter 19

Adlam, E. (2022). *The problem of temporal non-locality*. FQXi Essay Contest Winner.

Carroll, S. (2019). *Something Deeply Hidden: Quantum Worlds and the Emergence of Spacetime*. Dutton.

Descartes, R. *Meditations on First Philosophy*. Translated by John Cottingham. Cambridge University Press, 1996.

Goff, P. (2019). *Galileo's Error: Foundations for a New Science of Consciousness*. Pantheon Books.

Hoffman, D. D. (2019). *The Case Against Reality: Why Evolution Hid the Truth from Our Eyes*. W. W. Norton & Company.

Kastrup, B. (2019). *The Idea of the World: A Multidisciplinary Argument for the Mental Nature of Reality*. John Hunt Publishing.

LaBerge, S. (1985). *Lucid Dreaming*. Ballantine Books.

Plato. *The Republic*. Translated by G.M.A. Grube, revised by C.D.C. Reeve. Hackett Publishing, 1992.

Schopenhauer, A. *The World as Will and Representation*. Translated by E.F.J. Payne. Dover Publications, 1969.

Voss, U., Holzmann, R., Tuin, I., & Hobson, J. A. (2009). Lucid dreaming: A state of consciousness with features of both waking and dreaming. *Sleep, 32*(9), 1191–1200.

References for Chapter 20

Bem, D. J., & Honorton, C. (1994). Does Psi exist? Replicable evidence for an anomalous process of information transfer. *Psychological Bulletin*, 115(1), 4–18.

Chalmers, D. J. (1996). *The Conscious Mind: In Search of a Fundamental Theory*. Oxford University Press.

Greyson, B. (2000). Near-death experiences. In E. Cardeña, S. J. Lynn, & S. C. Krippner (Eds.), *Varieties of Anomalous Experience: Examining the Scientific Evidence* (pp. 315–352). American Psychological Association.

Griffiths, R. R., Richards, W. A., McCann, U., & Jesse, R. (2006). Psilocybin can occasion mystical-type experiences having substantial and sustained personal meaning and spiritual significance. *Psychopharmacology*, 187(3), 268–283.

Kim, Y.-H., Yu, R., Kulik, S. P., Shih, Y., & Scully, M. O. (2000). A delayed choice quantum eraser. *Physical Review Letters*, 84(1), 1–5.

LaBerge, S. (1985). *Lucid Dreaming*. Ballantine Books.

Levin, M. (2020). Bioelectricity: A new path to control of growth and form. *BioEssays*, 42(6), e1900201.

Wallace, B. A. (2007). *Contemplative Science: Where Buddhism and Neuroscience Converge*. Columbia University Press.

Wheeler, J. A. (1978). The "past" and the "delayed-choice" double-slit experiment. In A. R. Marlow (Ed.), *Mathematical Foundations of Quantum Mechanics* (pp. 9–48). Academic Press.

References for Chapter 21

Gödel, K. (1931). *On formally undecidable propositions of Principia Mathematica and related systems*. Springer.

Hoffman, D. D. (2019). *The Case Against Reality: Why Evolution Hid the Truth from Our Eyes*. W. W. Norton & Company.

Kim, Y.-H., Yu, R., Kulik, S. P., Shih, Y., & Scully, M. O. (2000). A delayed choice quantum eraser. *Physical Review Letters, 84*(1), 1–5.

LaBerge, S. (1985). *Lucid Dreaming*. Ballantine Books.

Livio, M. (2002). *The Golden Ratio: The Story of Phi, the World's Most Astonishing Number*. Broadway Books.

Recommended Reading

- Barrett, L. F. (2017). *How Emotions Are Made: The Secret Life of the Brain*. Houghton Mifflin Harcourt. *(Supports constructed emotion theory integration.)*
- Carroll, S. (2019). *Something Deeply Hidden: Quantum Worlds and the Emergence of Spacetime*. Dutton.
- Chalmers, D. J. (1996). *The Conscious Mind: In Search of a Fundamental Theory*. Oxford University Press.
- Gödel, K. (1931). *On Formally Undecidable Propositions of Principia Mathematica and Related Systems.*
(Foundational work referenced indirectly for limits of closed system explanations.)
- Harris, S. (2014). *Waking Up: A Guide to Spirituality Without Religion*. Simon & Schuster.
- Hoffman, D. D., Prakash, C. (2014). *Objects of Consciousness. Frontiers in Psychology. (Hoffman's more technical work behind interface theory.)*
- Hoffman, D. D. (2019). *The Case Against Reality: Why Evolution Hid the Truth from Our Eyes*. W.W. Norton & Company.
- James, W. (1890). *The Principles of Psychology*. Henry Holt and Company.
- Kastrup, B. (2019). *The Idea of the World: A Multidisciplinary Argument for the Mental Nature of Reality*. John Hunt Publishing.
- Metzinger, T. (2009). *The Ego Tunnel: The Science of the Mind and the Myth of the Self*. Basic Books.
- Nagel, T. (2012). *Mind and Cosmos: Why the Materialist Neo-Darwinian Conception of Nature Is Almost Certainly False.* Oxford University Press.
- Noë, A. (2009). *Out of Our Heads: Why You Are Not Your Brain, and Other Lessons from the Biology of Consciousness*. Hill and Wang.

- Radin, D. (2006). *Entangled Minds: Extrasensory Experiences in a Quantum Reality*. Simon & Schuster.
- Revonsuo, A. (2009). *Consciousness: The Science of Subjectivity*. Psychology Press. *(Deep work on dream studies and virtual reality models of mind.)*
- Spira, R. (2017). *Being Aware of Being Aware (The Essence of Meditation Series, Vol. 1)*. Sahaja Publications.
- Targ, R. (2012). *The Reality of ESP: A Physicist's Proof of Psychic Abilities*. Quest Books.
- Tononi, G. (2012). *Phi: A Voyage from the Brain to the Soul*. Pantheon Books.

Glossary of Terms

Awareness
The fundamental, formless presence in which all experience arises. Not something you do, but what you are. It is constant, unchanging, and timeless, the true subject behind all perceptions, thoughts, and identities.

Binding Problem
A challenge in neuroscience that asks how separate features of perception, like color, motion, sound, and shape, processed in distinct brain regions and at different times, are unified into a single, coherent experience. While some theories suggest synchrony of neural firing or attentional focus, no explanation accounts for how these elements are subjectively experienced as a whole. CDIT reframes the issue entirely: rather than being stitched together by the brain, unified experience is rendered whole within awareness. Perceived separateness is part of the dream interface, not a fragmentation of consciousness that must be reassembled.

Block Universe
A concept from modern physics (particularly relativity) that views time as a static dimension. Past, present, and future coexist within a four-dimensional spacetime "block," supporting CDIT's view that linear time is an illusion rendered within awareness.

CDIT (Cognitive Dream Interface Theory)
A metaphysical framework proposing that reality is a dreamlike simulation arising in universal awareness, where space, time, self, and thought are rendered by a symbolic interface. Brains

are not creators of consciousness but localizing elements within the dream, helping simulate narrative structure and autonomy.

Coherence Pressure
The tendency of dream narratives, including waking life, to maintain consistency and plausibility. This principle accounts for the appearance of physical laws, memory continuity, and identity stability in the dream interface.

Coherent Emergence
A concept in CDIT explaining how specific characters or scenes appear not randomly, but as structurally consistent elements within the dream. Identity arises with the rendered scene, not before it.

Constructed Emotion Theory
Lisa Feldman Barrett's theory proposing that emotions are not hardwired programs but constructed interpretations based on bodily input, prediction, and context. Supports CDIT's view that emotion is part of the symbolic interface, not an essential interior truth.

Dissociation (Kastrup's Model)
Philosopher Bernardo Kastrup's model of consciousness suggests that apparent individual minds are dissociated alters within a single universal consciousness. While CDIT shares the nondual foundation, it diverges by avoiding dissociation metaphors, favoring the view that all selves are symbolic localizations, not separations, of awareness.
(Kastrup, 2019, *The Idea of the World*)

Dream Analogy
A key metaphor in the manuscript comparing waking reality to a dream, not to deny its vividness, but to highlight its status

as symbolic, immersive, and generated within awareness itself.

Dreamer vs. Dream Character

In CDIT, the dreamer is awareness itself; timeless, formless, and unchanging. The dream character is the localized self rendered within the symbolic interface. The character feels autonomous but is ultimately a viewpoint rendered within the dream.

Dreaming Intelligence

The implicit structure and logic by which awareness spontaneously renders coherent worlds, scenes, characters, and narratives. Not a conscious intention or planning agent, but a naturally occurring capacity for symbolic organization, like gravity's capacity to attract. It explains the appearance of meaningful form in dreams and waking life without requiring a separate designer.

Emergence (Critique of)

In CDIT, emergence is reframed as a perceptual illusion. Just as the liquidity of water cannot be precisely located in a specific number of H_2O molecules, or an anthill in a grain of sand, consciousness does not emerge from neurons. All emergence is rendered after-the-fact, as logical backfill within the dream. There is no true moment of origin, only appearances arising in awareness.

Fibonacci Sequence/Golden Ratio

A numerical series in which each number is the sum of the two preceding ones (e.g., 0, 1, 1, 2, 3, 5, 8...). Frequently observed in biological forms, patterns in nature, and aesthetics, it serves in CDIT as symbolic evidence of inherent order and symmetry emerging from the dreamer's creative rendering.

Ganzfeld Effect
A psychological phenomenon where unstructured, uniform sensory input leads to hallucinations or vivid internal imagery. Used in CDIT to illustrate that when input fades, awareness begins to spontaneously render experience from within.

Gödel's Incompleteness Theorem
A foundational result in mathematical logic showing that any formal system complex enough to describe arithmetic will contain true statements that cannot be proven within the system. CDIT uses this to argue that materialist or simulation-based models (as closed systems) cannot fully explain reality, but awareness, being outside the system, can.

Idealism
The philosophical view that consciousness or awareness is the fundamental ground of reality, and that the material world emerges within it, not the other way around.

Interface Icon
A symbol or perception rendered in the dream interface, like time, space, or color, used to navigate the dream, not to depict its underlying nature. Borrowed and expanded from Donald Hoffman's "user interface theory."

Localization (of Awareness)
The focusing of awareness into a specific dream character or perspective, giving rise to the illusion of individuality. This process simulates separation without true division.

Lucid Dreaming
A state in which the dreamer becomes aware they are dreaming. In CDIT, this metaphor extends to waking life: when awareness recognizes itself within the waking dream, lucidity arises.

Narrative Self
The autobiographical sense of self constructed by memory, language, and internal narration. In CDIT, the self is not an entity but a story, useful, coherent, but ultimately illusory.

Nonduality
The recognition that apparent separation between self and other, subject and object, is illusory. All phenomena arise within a single awareness, which is never actually divided.

Observer Effect
In quantum physics, the principle that measurement affects the system being measured. Used metaphorically in CDIT to suggest that awareness plays a fundamental role in rendering reality.

Recursive Dreaming
The phenomenon of dreams within dreams, or layered states of experience. Used in CDIT to illustrate how awareness can simulate nested realities without fragmentation.

Spontaneous Rendering
The process by which awareness generates coherent symbolic content, as in dreams, without prior templates or instructions. Supports CDIT's view that form arises naturally from the potential of awareness.

Symbolic Interface
The perceptual, cognitive, and sensory environment we call reality. According to CDIT, this interface is not a reflection of objective reality, but a functional illusion rendered for narrative and experiential coherence.

Transceiver Theory (of the Brain)
The theory that the brain does not produce consciousness but modulates or "tunes into" an existing field of awareness. CDIT adapts this to say the brain is a rendered component of the dream, a symbol of localization, not a source.

Appendix

Statement of Originality

Cognitive Dream Interface Theory (CDIT) is an original philosophical model developed through independent synthesis. While drawing conceptual influence from thinkers such as Donald Hoffman (user interface theory), Bernardo Kastrup (analytic idealism), Emily Adlam (temporal non-locality), and Rupert Spira (nondual awareness), CDIT proposes a distinct metaphysical framework. It unites elements of the dream analogy, symbolic interface rendering, and awareness-based ontology into a novel theory of reality. Though informed by existing scientific and philosophical currents, CDIT is not derived from or identical to any previously published framework. All conclusions, narrative structures, and theoretical integrations herein are the product of original research, experiential inquiry, and creative development.

Additionally, this manuscript was developed with the assistance of AI-based tools for drafting, editing, research integration, and structural refinement. While the core ideas, conceptual frameworks, and overall direction are the author's own, portions of the text were initially generated by AI under the author's guidance, then reviewed, edited, and approved to ensure accuracy, coherence, and philosophical clarity.

ESSENTIA

Essentia Books, a collaboration between Collective Ink Ltd. and Essentia Foundation, publishes rigorous scholarly work relevant to metaphysical idealism, the notion that reality is essentially mental in nature. For more information on modern idealism, please visit www.essentiafoundation.org.

Recent Bestsellers from Essentia Books Are:

Irreducible
Federico Faggin
The inventor of the microprocessor explores
the fundamental nature of consciousness.
Paperback: 978-1-80341-509-3 ebook: 978-1-80341-510-9

The Sapient Cosmos
James B. Glattfelder
Novel paradigms in physics and philosophy are
uncovering radical new visions for the ultimate
nature of reality and consciousness.
Paperback: 978-1-80341-404-1 ebook: 978-1-80341-405-8

Children's Unexplained Experiences in a Post Materialist World
Donna Maria Thomas
Paperback: 978-1-80341-084-5 ebook: 978-1-80341-085-2

Mind, Matter and Life
Massimiliano Sorrentino and Daniela Panighetti
Why the mind-matter problem could invalidate the neo-Darwinian materialist conception of the world.
Paperback: 978-1-80341-530-7 ebook: 978-1-80341-613-7

What Lies Beyond
Matt Colborn
Paranormal experiences — why they matter and
how they can help make a better tomorrow.
Paperback: 978-1-80341-829-2 ebook: 978-1-80341-836-0

In Pursuit of the Inconceivable

Peter Guy Jones

A comprehensive explanation of the relationship between formal scholastic metaphysics and the philosophical foundation of mysticism and the Perennial philosophy.

Paperback: 978-1-80341-906-0 ebook: 978-1-80341-921-3

Find more titles at

www.collectiveinkbooks.com/essentia-books